ELEMENTARY ELECTROCHEMISTRY

ELEMENTARY ELECTROCHEMISTRY

A. R. DENARO, M.Sc., Ph.D., F.R.I.C.

Liverpool Polytechnic

BUTTERWORTHS

LONDON - BOSTON

Sydney · Wellington · Durban · Toronto

THE BUTTERWORTH GROUP

ENGLAND

Butterworth & Co (Publishers) Ltd
London: 88 Kingsway, WC2B 6AB

AUSTRALIA

Butterworths Pty Ltd
Sydney: 586 Pacific Highway,
    Chatswood, NSW 2067
Also at Melbourne, Brisbane, Adelaide
    and Perth

SOUTH AFRICA

Butterworth & Co (South Africa) (Pty) Ltd
Durban: 152–154 Gale Street

NEW ZEALAND

Butterworths of New Zealand Ltd
Wellington: 26–28 Waring Taylor Street, 1

CANADA

Butterworth & Co (Canada) Ltd
Toronto: 2265 Midland Avenue,
    Scarborough, Ontario, M1P 4S1

USA

Butterworths (Publishers) Inc
Boston: 19 Cummings Park,
    Woburn, Mass. 01801

First Published 1965
Second Impression 1968
Second Edition 1971
Second Impression 1976

©
Butterworth & Co (Publishers) Ltd,
1971

Suggested U.D.C. number: 541·13

ISBN 0 408 70071 8

Printed in Great Britain by offset lithography by
Billing & Sons Ltd, Guildford, London and Worcester

PREFACE

THIS BOOK is intended for students reading electrochemistry for the first time after G.C.E. 'A' level or Ordinary National Certificate in Science. The first edition was aimed primarily at students who proposed to take the Part I examination of the Royal Institute of Chemistry but in this edition the last chapter has been extended to give a more quantitative treatment of the kinetics of electrode processes. The scope of the book thus goes beyond the Part I Grad.R.I.C. syllabus and should provide an introduction to the basic ideas of electrode processes required in more advanced studies. In this way the book covers the electrochemistry required for the Part I Grad.R.I.C. examination and for Higher National Certificate in chemistry and should prove a useful introduction to the subject for students on degree courses.

It is assumed that the student's knowledge of thermodynamics and kinetics will advance in step with his study of electrochemistry and thermodynamic and kinetic relationships have been quoted where necessary. A qualitative introduction to the activity concept is provided, as it may happen that this concept is required in electrochemistry before it has been treated in a thermodynamics course.

In this second edition the opportunity has been taken to convert the units of quantities and the nomenclature to those recommended by the Système International d'Unités (SI). As the SI incorporates the International Sign Convention of Electrode Potentials, which is now more widely accepted, this has been incorporated in the text rather than being relegated to an appendix as in the first edition.

Some suggestions for further reading have been given after the last chapter but the books quoted have not yet appeared in editions which conform to the SI. By the time the student wishes to extend his reading however, it is hoped that his grasp of the fundamental principles of electrochemistry is such as to render any transition between systems of units a trivial problem.

A. R. D.

CONTENTS

CONTENTS

1

INTRODUCTION

THE ORIGINS OF ELECTROCHEMISTRY

IT WAS FIRST observed by Nicholson and Carlisle in 1800 that when two inert wires, for example platinum, connected to the poles of a voltaic cell or battery, were immersed in a dilute aqueous solution of sulphuric acid, a current flowed round the circuit. At the same time bubbles of hydrogen were evolved at one wire and bubbles of oxygen at the other. Obviously some breakdown of the solution was occurring under the influence of the electric current. With other systems, similar effects could be observed: evolution of gases, deposition of material on the wires in the solution or even dissolution of the wires themselves. These are all examples of the phenomenon termed *electrolysis*.

The first quantitative studies of electrolysis were carried out by Faraday in 1833, and it was he who first used the nomenclature which is still employed today. The wires, or *electrodes*, which were immersed in the solution were called the *anode* and the *cathode*, respectively. The anode may be defined as the electrode at which negative electricity (electrons) *leaves* the solution, and the cathode as the electrode at which negative electricity *enters* the solution. These definitions are universal and apply to any situation. The solution which carries the current is termed the *electrolyte*, that in the immediate neighbourhood of the anode being the *anolyte* and that around the cathode being the *catholyte*.

To account for the passage of a current through the electrolyte, Faraday assumed that the flow of electricity was due to the movement of charged particles which he called *ions*. Those ions which move towards the anode are called *anions* and those which move towards the cathode are called *cations*.

It will be seen from these definitions that, if the cathode is the lectrode where negative electricity enters the solution, it must be

1

connected to a source of negative electricity. That is to say, it will be connected to the negative pole of the battery and will itself be negatively charged. The cations which travel towards the cathode must therefore be positively charged, as it is the charge supplied by the battery which is attracting the ions to the electrodes. Conversely, anions carry a negative charge. On reaching the electrodes, the ions were assumed to have their charges neutralised when they formed normal atoms or molecules.

ELECTRONIC AND ELECTROLYTIC CONDUCTORS

From a consideration of an electrolysis there emerge two classes of conductors of electricity:

(1) *Electronic conductors*—These are conductors where no transfer of material occurs when a current is passed. The wires connecting the electrodes to the battery fall into this class, and we now interpret an electric current in these cases as a flow of electrons along the wire.

(2) *Electrolytic conductors*—These are conductors where there is a transfer of material associated with the passage of an electric current. The solution which is being electrolysed provides such an example. The current is carried by the ions which travel through the solution towards the electrodes, and there is a resultant change of concentration throughout the solution.

Electrolytic conductors may be subdivided into two sections:

- (*a*) pure substances, e.g. fused potassium hydroxide, fused sodium chloride;
- (*b*) solutions, e.g. of acids, bases and salts in water or other solvents.

It is this latter category which has been most extensively studied and with which we shall be mostly concerned in this book.

It is worth emphasising a further distinction between electronic and electrolytic conductors at this stage. Whereas current in an electronic conductor is associated with the flow of negative charges in only one direction, it must always be remembered that in electrolytic conductors current is associated with the transfer of both negative and positive charges in opposite directions.

SYSTEMS OF UNITS

We must now establish the units in which various chemical and electrochemical quantities are measured. In this connection it may be useful to first consider the units of mechanical quantities.

The fundamental dimensions in which mechanical quantities may be expressed are chosen as mass, length and time. If mass is measured in grammes, length in centimetres and time in seconds, the resulting units are CGS (centimetre-gramme-second) units. Alternatively, if mass is measured in kilogrammes, length in metres and time in seconds, the resulting units are MKS(metre-kilogramme-second) units.

In general, quantities of interest to chemists have been expressed in the CGS system but as various fields of study have developed within chemistry, additions have been made to the CGS system to express results pertinent to a particular field. Thus although thermochemists, photochemists, spectroscopists and nuclear chemists all wish to speak of energy they may use different units to express this physical quantity. As a result, there is in chemistry, a multiplicity of units for energy. The same may be said of other branches of science such as electricity and magnetism and in crossing the boundaries between chemistry and engineering the situation is even more complex.

In order to simplify the position an attempt is being made currently to introduce one system of units which shall be agreed internationally and in which there is, for example, only one unit of energy. This system is known as the Système International d'Unités which is usually abbreviated SI. SI units are based on the MKS system and we shall start by considering mechanical quantities.

The unit of velocity in the MKS system is, of course, the $m\ s^{-1}$ and the unit of acceleration is the $m\ s^{-2}$.

Force, F, may be defined in terms of acceleration by the equation

$$F = ma \tag{1.1}$$

where m is the mass of the body undergoing acceleration. As unit mass is 1 kg, unit force is given by equation (1.1) as

$$F = 1 \text{ kg} \times 1 \text{ m s}^{-2} = 1 \text{ kg m s}^{-2}$$

The MKS unit of force is the kg m s^{-2} which is called the *newton*, N.

Work or energy, w, is related to force by the equation

$$w = Fl \qquad (1.2)$$

where l is the distance through which the force acts. The unit of work is thus given by equation (1.2) as

$$w = 1 \text{ N} \times 1 \text{ m} = 1 \text{ N m}$$

The MKS unit of energy is the N m which is called the *joule*, J. This is the work done when a force of one newton acts through a distance of one metre.

Power, P, is related to energy by the equation

$$P = \frac{w}{t} \qquad (1.3)$$

so that unit power is developed when unit work is done in unit time. Equation (1.3) thus gives unit power as

$$P = \frac{1 \text{ J}}{1 \text{ s}} = 1 \text{ J s}^{-1}$$

The MKS unit of power is thus the J s^{-1} which is called the *watt*, W.

Hence, we have the relationships

$$1 \text{ kg} \times 1 \text{ m s}^{-2} = 1 \text{ N}$$
$$1 \text{ N} \times 1 \text{ m} = 1 \text{ J}$$
$$1 \text{ J} \times 1 \text{ s}^{-1} = 1 \text{ W}$$

When systems of mechanical units are extended to cover electrical and magnetic quantities, a fourth quantity must be defined in addition to mass, length and time. There are several choices available but the SI has electric current as the fourth fundamental quantity. The basic SI unit of current is defined as the ampère, A, and hence all electrical and magnetic quantities can be expressed in the dimensions mass, length, time and current, so that the SI is a MKSA(metre-kilogramme-second-ampère) system. The SI ampère is defined in such a way that it is identical with the established ampère which has always been the practical unit of current.

Once the basic unit of current has been taken as the ampère, unit charge can be defined as the amount of charge transferred per unit time by unit current. The relationship between an amount of charge Q, passed by a current I, flowing for a time t, is

$$Q = It \qquad (1.4)$$

Unit charge is thus given by

$$Q = 1\,\mathrm{A} \times 1\,\mathrm{s} = 1\,\mathrm{A\,s}$$

This amount of charge is called the *coulomb*, C.

The relationship between potential difference U, the current flowing I, and the power P, is

$$U = \frac{P}{I} \qquad (1.5)$$

Unit potential difference is thus given by

$$U = \frac{1\,\mathrm{W}}{1\,\mathrm{A}} = 1\,\mathrm{W\,A^{-1}}$$

The SI unit of potential difference is thus the $\mathrm{W\,A^{-1}}$ which is called a *volt*, V. An alternative statement may be made by multiplying the numerator and denominator of the right-hand side of equation (1.5) by time

$$U = \frac{Pt}{It} = \frac{w}{Q} \qquad (1.6)$$

Unit potential difference could thus be equally well defined such that 1 J of work is performed when 1 C of charge is transferred through unit potential difference. Thus from equation (1.6)

$$1\,\mathrm{V} = \frac{1\,\mathrm{J}}{1\,\mathrm{C}} = 1\,\mathrm{J\,C^{-1}} = 1\,\mathrm{J\,A^{-1}\,s^{-1}} = 1\,\mathrm{W\,A^{-1}}$$

The definition of unit resistance is provided by Ohm's law which relates potential difference U, current I, and resistance R, by the equation

$$U = IR \qquad (1.7)$$

Unit resistance is such that when unit potential difference is applied to it, unit current flows. Equation (1.7) thus gives unit resistance as

$$R = \frac{1 \text{ V}}{1 \text{ A}} = 1 \text{ V A}^{-1}$$

The SI unit of resistance is thus the V A^{-1} which is called the *ohm*, Ω.

As pointed out above the SI is based on the MKSA system but it includes two other fundamental quantities in addition to mass, length, time and current. These are thermodynamic temperature and luminous intensity. This latter quantity has little relevance to chemistry and will therefore be neglected. It is hoped that the use of a seventh fundamental quantity, amount of substance, will soon be confirmed and for the purposes of this book it will be regarded as a fundamental quantity in the SI.

The basic unit of thermodynamic temperature is the *kelvin*, K. This was formerly denoted by the symbol °K or sometimes deg. Only the symbol K will now be used.

The basic unit of amount of substance is the *mole*, mol. It is defined as 'the amount of substance which contains as many elementary units as there are atoms in 0·012 kg of carbon-12. The elementary unit must be specified and may be an atom, a molecule, an ion, a radical, an electron, a photon, etc., or a specified group of such entities.' Thus a mole of H_2SO_4 means $6·02 \times 10^{23}$ units of H_2SO_4 and a mole of $\frac{1}{2}H_2SO_4$ means $6·02 \times 10^{23}$ units of $\frac{1}{2}H_2SO_4$. The term, 'a mole of sulphuric acid' could thus be somewhat ambiguous unless the meaning of sulphuric acid was specified. If the term mole is used without specification of the elementary unit of substance, it should be taken that the elementary unit is the usual formula of the substance. Thus, a mole of magnesium sulphate should be interpreted as meaning $6·02 \times 10^{23}$ units of $MgSO_4$. The result of the definition of the unit mol is simply to confirm the use which this quantity has always had in chemistry.

INTRODUCTION

The SI permits the use of some prefixes to denote decimal fractions and multiples of the basic SI units. Some of these prefixes are given below.

Factor	Prefix	Symbol	Factor	Prefix	Symbol
10^{-1}	deci	d	10^1	deka	da
10^{-2}	centi	c	10^2	hecto	h
10^{-3}	milli	m	10^3	kilo	k
10^{-6}	micro	μ	10^6	mega	M
10^{-9}	nano	n	10^9	giga	G
10^{-12}	pico	p	10^{12}	tera	T

Thus a quantity of $2 \cdot 3 \times 10^9 \, \Omega$ could be written $2 \cdot 3 \, G\Omega$ or a quantity $1 \cdot 6 \times 10^{-6} \, V$ could be written $1 \cdot 6 \, \mu V$.

By virtue of the way in which various quantities are defined in the SI there are some consequences which are worth pointing out. In the SI, volume is considered to be (length)³. The basic SI unit of volume is thus the m^3. Smaller volumes can be expressed in terms of decimal fractions of the m^3, such as cm^3 or dm^3. In the SI, volumes are not expressed in litres or millilitres as the litre is not an SI unit. Concentrations of solutions which are the amount of substance divided by volume will thus be expressed in units such as $mol \, dm^{-3}$ or $mol \, cm^{-3}$ or such other similar units as are appropriate.

The term molar is not used to describe the concentration of a solution but is only used as an adjective to indicate a quantity which relates to one mole of material as in molar mass (the mass of one mole) or molar volume (the volume of one mole).

The only permitted unit for amount of substance is the mol. This renders units such as equivalents obsolete and they will not be used in this book.

The molar mass of a substance is the mass divided by the amount of substance and thus has the basic units $kg \, mol^{-1}$. The relative molecular mass is a dimensionless quantity defined as the ratio of the average mass per molecule divided by 1/12 of the mass of an atom of carbon-12. The term relative molecular mass thus replaces the older term 'molecular weight'.

Methods of expression in tables and graphs

Whilst the method of expressing results in the form of tables or graphs has no relation to the system of units used this seems to be an appropriate place to clarify the situation. Frequently a column of figures giving the values of (say) an electric current under various conditions is headed 'Current × 10^3, A' and the reader is not sure whether the figures given should be multiplied or divided by 10^3 to obtain the actual values of the current in ampères. In this book the situation will be resolved in the following way.

The value of a physical quantity is equal to the product of a pure number and a unit, e.g.,

$$c = 3 \cdot 6 \times 10^{-3} \text{ mol dm}^{-3}$$

This equation can be rearranged to give

$$\frac{c}{\text{mol dm}^{-3}} = 3 \cdot 6 \times 10^{-3}$$

or

$$\frac{c}{10^{-3} \text{ mol dm}^{-3}} = 3 \cdot 6$$

To avoid repetition of the unit symbol it is common practice to tabulate information in the form of pure numbers. It follows that a column in which the pure number 3·6 appears as corresponding to a concentration of $3 \cdot 6 \times 10^{-3}$ mol dm^{-3} it should carry the heading $c/(10^{-3}$ mol dm$^{-3})$. The same considerations apply to the labelling of the axes of graphs where pure numbers are placed at points along the axes.

FARADAY'S LAWS OF ELECTROLYSIS

The results of Faraday's investigations into the phenomenon of electrolysis may be summarised in his two laws of electrolysis.

(1) The amount of primary product formed at an electrode by electrolysis is directly proportional to the quantity of electricity passed.

(2) The amounts of various primary products formed at an electrode by the same quantity of electricity are proportional to their relative molecular or relative atomic masses divided

by the change in their charge numbers for the electrode process.

From the first law it follows that the same amount of a given substance is deposited or dissolved at an electrode by a given quantity of electricity. From the second law it follows that the amount of a substance deposited or dissolved at an electrode by a given quantity of electricity is proportional to M_r/z (or A_r/z) where M_r is the relative molecular mass (or A_r is the relative atomic mass) of the substance and z is the change in its charge number which is associated with the electrode process. Thus for the formation of $1/z$ mol of any substance the same quantity of electricity is required. This is found to be 96 487 C. It will be realised that the electric charge carried by 1 mol of an ion is equal to the product of its charge number and the Faraday constant, F, which has the value of 96 487 C mol^{-1}.

These statements will probably be clarified by the consideration of two examples. In the electrolysis of copper (II) sulphate between copper electrodes, the process occurring at the cathode is the deposition of copper

$$Cu^{2+} + 2e \rightarrow Cu$$

The change in the charge number of copper in this process is 2 and hence 96 487 C of electricity will deposit $\frac{1}{2}$ mol of Cu.

In the electrolysis of potassium ferrocyanide between platinum electrodes, the process occurring at the anode is the oxidation of ferrocyanide ions to ferricyanide ions.

$$Fe(CN)_6{}^{4-} \rightarrow Fe(CN)_6{}^{3-} + e$$

Here, the change in the charge number is 1 and hence 96 487 C of electricity will lead to the formation of 1 mol of $Fe(CN)_6{}^{3-}$.

COULOMETERS

It is frequently necessary to know the quantity of electricity that has passed through a circuit. This could be done by measuring the current which passes for a known time, but this method is unsatisfactory for fluctuating currents. The application of Faraday's laws provides a method of overcoming the problem. As the amount of any substance deposited or dissolved at an electrode is proportional to the quantity of electricity passed, it is only necessary to incorporate an electrolytic cell in the circuit and to determine the amount of reaction which has occurred during the passage of the current. Such devices are termed *coulometers*. There are several types, the more important being mentioned below.

(a) The copper coulometer

The copper coulometer consists of a copper anode and a copper cathode immersed in a solution of copper sulphate usually containing sulphuric acid and alcohol. On the passage of current, the copper anode goes into solution and metallic copper is deposited on the cathode. The alcohol in the solution inhibits the oxidation of the freshly deposited copper on the cathode. The cathode is weighed before and after electrolysis and the quantity of electricity passed can be calculated from the increase in weight of the cathode.

The copper coulometer is a robust device of moderate accuracy suitable for measuring quantities of electricity up to about 0·1 faraday.

(b) The iodine coulometer

The iodine coulometer consists of separate anode and cathode compartments, joined by a narrow tube which eliminates diffusion between the anolyte and the catholyte. The electrodes are of platinum–iridium alloy. The anolyte is a concentrated solution of potassium iodide and the catholyte is a standard solution of iodine in potassium iodide. The intervening electrolyte between the anolyte and the catholyte is a 10 per cent solution of potassium iodide.

When current is passed, the iodide ions around the anode give up electrons and are converted to iodine and, at the cathode, the iodine already existing is reduced to iodide ions by the electrons supplied by the cathode. The quantity of electricity which has been passed is calculated from the change in the iodine concentration around the electrodes.

The iodine coulometer has a greater accuracy than the copper coulometer and can, under certain circumstances, measure amounts of electricity as low as 25 C to about 0·1 per cent.

(c) *The gas coulometer*

In the gas coulometer, hydrogen and oxygen are generated at a platinum cathode and anode, respectively, by the electrolysis of an aqueous solution of a suitable electrolyte. The volume of the total amount of gas is measured at a known temperature and pressure. As a small weight of gas occupies a large volume, the gas coulometer is useful for measuring very small quantities of electricity down to about 5 C.

(d) *The silver coulometer*

This is probably the most accurate coulometer. The cathode is a platinum crucible containing a solution of pure silver nitrate. On electrolysis, metallic silver is deposited on the platinum crucible, the depletion of the electrolyte being made good by the dissolution of a silver anode suspended in the solution. The anode is surrounded by a porous pot which catches any solid particles falling from the anode, preventing them from settling in the platinum crucible. The crucible is weighed before and after electrolysis and the quantity of electricity passed is calculated from the weight of silver deposited.

2

ELECTROLYTIC CONDUCTION

RESISTIVITY AND CONDUCTIVITY

THE RESISTANCE, R, of any electrical conductor is proportional to its length, l, and inversely proportional to its cross-sectional area, a

$$R \propto \frac{l}{a}$$

It is therefore possible to write

$$R = \frac{\rho l}{a} \tag{2.1}$$

where ρ is a proportionality constant known as the *resistivity* of the conductor. As R is a resistance, l is a length and a is an area it can be seen from equation (2.1) that ρ must have the units of (resistance) (length). For example, a piece of copper of length $0 \cdot 2$ m and cross-sectional area 10^{-4} m² has a resistance of $3 \cdot 45 \times 10^{-5} \, \Omega$. From equation (2.1)

$$\rho = \frac{Ra}{l}$$

and substituting the above values, the resistivity of copper is given by

$$\rho = \frac{3 \cdot 45 \times 10^{-5} \, \Omega \times 10^{-4} \, \text{m}^2}{0 \cdot 2 \, \text{m}}$$

or

$$\rho = 1 \cdot 725 \times 10^{-8} \, \Omega \, \text{m}$$

In electrolytic conduction we are more concerned with conductance than resistance, and as a material is a better conductor the lower its resistance, conductance is related in a reciprocal fashion to resistance. We therefore define conductivity κ, as the reciprocal of resistivity

$$\kappa = \frac{1}{\rho} \qquad (2.2)$$

from which relationship the units of conductivity are seen to be (resistance)$^{-1}$ (length)$^{-1}$. *Table 1* gives some values of conductivities for various materials.

Table 1

CONDUCTIVITIES OF MATERIALS AT 25°C

Material	Conductivity $\overline{\Omega^{-1}cm^{-1}}$
Silver	$6·33 \times 10^5$
Copper	$5·80 \times 10^5$
Fused sodium chloride	$3·3$
0·1 mol dm^{-3} KCl, aq	$1·29 \times 10^{-2}$
0·1 mol dm^{-3} NaOH, aq	$2·21 \times 10^{-2}$
0·1 mol dm^{-3} CH$_3$COOH, aq	$5·20 \times 10^{-4}$
Water	$4·0 \ \times 10^{-8}$
Sulphur	$2·5 \ \times 10^{-16}$

(After *Handbook of Chemistry and Physics*, 37th ed., Chemical Rubber Publ. Co., 1955)

The conductivity may be expressed in terms of resistance by combining equations (2.1) and (2.2) to give

$$\kappa = \frac{l}{Ra} \qquad (2.3)$$

According to Ohm's law

$$R = \frac{U}{I}$$

and hence

$$\kappa = \frac{Il}{Ua}$$

or

$$\kappa = \frac{I/a}{U/l}$$

I/a is the current per unit cross-sectional area. This is known as the current density and represented by the symbol j. U/l is the potential fall per unit length which is known as the potential gradient or electric field intensity, E. Thus

$$\kappa = \frac{j}{E} \qquad (2.4)$$

and the conductivity may be regarded as the current passing across unit area under unit potential gradient.

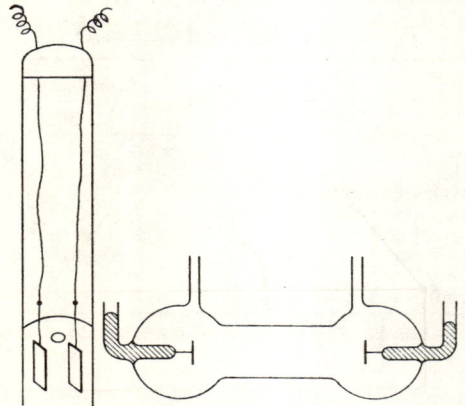

Figure 1. Conductance cells

MEASUREMENT OF THE CONDUCTANCE OF ELECTROLYTES

In measuring the conductance of an electrolyte, the property which is actually determined is the resistance, the conductivity being obtained from equation (2.3).

The electrolyte solution is accommodated in a *conductance cell* which comprises one arm of a Wheatstone bridge circuit. Two examples of conductance cells are shown in *Figure 1*. They usually consist of glass vessels containing two electrodes a fixed distance apart and it is the resistance of that volume of solution contained between the electrodes which is measured. It can thus be understood that when the resistance of a weakly conducting solution is to be determined, it is preferable to have electrodes of large cross-sectional area separated by only a small distance. This will reduce the resistance of the electrolyte contained between the electrodes to a value which is conducive to easy measurement [see eqn. (2.1)]. With a more strongly conducting solution the electrodes may be smaller and farther apart.

It is impossible to use d.c. when measuring the resistance of an electrolyte solution with inert electrodes, as electrolysis would

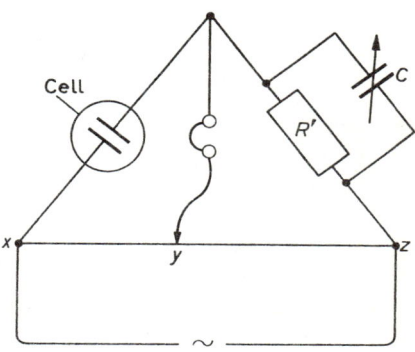

Figure 2. Wheatstone Bridge circuit

then occur, the products accumulating at the electrodes and interfering with the measurement. For this reason a.c. of about 1000 Hz is used so that the small amount of electrolysis taking place in one half-cycle is completely reversed in the opposite half-cycle. In order to reduce any residual interfering effects to a minimum, platinised platinum electrodes are used in the conductance cells. These are platinum electrodes upon which more platinum in a very fine state of division has been electrolytically deposited. Some accurate measurements have been made in recent years using d.c. but special electrodes are required in

15

these cases, and the a.c. method is the standard one for general applications.

The principle of the complete Wheatstone bridge circuit is represented in *Figure 2*, the four arms of the bridge being the conductance cell, a variable resistance, R', and the two sections of the bridge wire, xy and yz. The resistance R' is set to a value which is of the order of the resistance of the cell, and the bridge is balanced by moving a sliding contact along the bridge wire xz. The detector in the circuit is a low-resistance earphone and the achievement of balance is indicated by the minimum intensity of sound in the earphone. As a.c. is used, it is necessary to balance not only the resistances of the bridge arms but also the capacitances. Provision is made for this by having a variable capacitance in parallel with the resistance R'. With balanced capacitances, the position of minimum signal in the earphone is more easily recognised.

If the balance point is at y, as shown in *Figure 2*, then the resistance, R, of the cell is obtained from the expression

$$\frac{R}{R'} = \frac{xy}{yz} \tag{2.5}$$

The instruments used in practice are rather more sophisticated than that illustrated in *Figure 2*, the bridge wire being replaced by variable resistors. The balance point is often detected by a 'magic eye' in such instruments.

Conductance varies markedly with temperature and the conductance cell should always be contained in a thermostat.

CELL CONSTANT

The conductivity of the solution is related to its resistance by eqn. (2.3). In conductance measurements, l may be taken as the distance between the electrodes of the cell and a as the area of the electrodes. For a given cell, l and a will be constant, and l/a is known as the *cell constant*. The cell constant may be measured directly but it is more usual to deduce its value by measuring in the cell the resistance of an electrolyte of accurately known resistivity. The commonest solutions used are those of potassium chloride.

ELECTROLYTIC CONDUCTION

Suppose a given solution of potassium chloride of conductivity κ_0 has a resistance R_0 in a given cell. Then, by eqn. (2.3)

$$\kappa_0 = \frac{l}{R_0 a}$$

and

$$l/a = \kappa_0 R_0$$

If R is the resistance of another solution in the same cell, then its conductivity, κ, is given by

$$\kappa = \frac{l}{Ra} = \frac{\kappa_0 R_0}{R} \tag{2.6}$$

MOLAR CONDUCTIVITY

The conductivities of electrolytes vary greatly with concentration. Comparison of the conductances of different electrolytes is more useful when some account is taken of concentration and for this reason a new term is defined.

Molar conductivity, Λ, is defined by

$$\Lambda = \kappa/c \tag{2.7}$$

The basic units of molar conductivity are $\Omega^{-1}\,m^2\,mol^{-1}$ as can be seen from the following example. The conductivity of potassium chloride solution at a concentration of $10^2\,mol\,m^{-3}$ is $1{\cdot}29\,\Omega^{-1}\,m^{-1}$ at $25\,°C$. The molar conductivity of this solution is thus given by

$$\Lambda = \frac{1{\cdot}29\,\Omega^{-1}\,m^{-1}}{10^2\,mol\,m^{-3}}$$

$$= 1{\cdot}29 \times 10^{-2}\,\Omega^{-1}\,m^2\,mol^{-1}$$

Although the above result is given in *basic* units there is, of course, no reason why it should not be expressed as

$$\Lambda = 129\,\Omega^{-1}\,cm^2\,mol^{-1}$$

the units $\Omega^{-1}\,cm^2\,mol^{-1}$ being equally acceptable.

A further point to be observed in the comparison of the molar conductivities of various electrolytes is that electric current is

17

the transfer of charge. It is thus preferable, in comparisons, to consider molar conductivities where the various amounts of substances comprising a mole carry the same number of charges. For example, in the comparison of the molar conductivities of sodium chloride and zinc sulphate it is better to specify a mole of sodium chloride as NaCl and to specify a mole of zinc sulphate as $\frac{1}{2}ZnSO_4$. If the moles of sodium chloride and zinc sulphate are specified in this way, a mole of each provides the same number of charges and the molar conductivities of each will be more comparable. This point is illustrated in *Table 2*.

VARIATION OF MOLAR CONDUCTIVITY WITH CONCENTRATION

Some data for various electrolytes are given in *Table 2*. These results show that, as the concentration decreases or the dilution

Table 2

MOLAR CONDUCTIVITIES OF AQUEOUS SOLUTIONS AT 25°C

Concentration mol dm^{-3}	Electrolyte						
	HCl	KCl	NaCl	AgNO$_3$	$\frac{1}{2}$ZnSO$_4$	$\frac{1}{2}$NiSO$_4$	HAc
0·0005	422·7	147·8	124·5	131·4	121·4	118·7	
0·001	421·4	147·0	123·7	130·5	114·5	113·1	48·63
0·005	415·8	143·6	120·7	127·2	95·5	93·2	22·80
0·01	412·0	141·3	118·5	124·8	84·9	82·7	16·20
0·02	407·2	138·2	115·8	121·4	74·2	72·3	11·57
0·05	399·1	133·4	111·1	115·2	61·2	59·2	7·36
0·10	391·3	129·0	106·7	109·1	52·6	50·8	5·20

(After *Handbook of Chemistry and Physics*, 37th ed., Chemical Rubber Publ. Co., 1955)

increases, the molar conductivity increases and tends towards a limiting value which is known as the limiting molar conductivity and is denoted by the symbol Λ^∞. This behaviour is generalised in *Figure 3*.

From extensive studies of the conductance of electrolytee, Kohlrausch (1900) was able to show that in some cases, for diluts solutions, the relationship between molar conductivity, Λ, and concentration, c, could be expressed by the empirical equation.

$$\Lambda = \Lambda^{\infty} - k\sqrt{c} \tag{2.8}$$

where k is a constant. This is the equation of a straight line of slope $-k$ and intercept Λ^{∞}, and the relationship permits Λ^{∞} to be determined by extrapolation to zero concentration in those cases where the equation is applicable.

In *Figure 4* values of Λ^{∞} are plotted against $\sqrt{c}$ for various

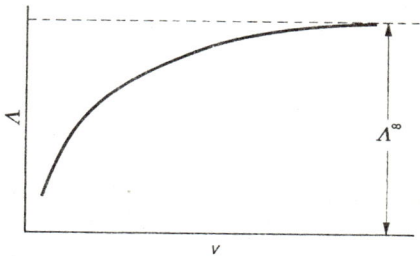

Figure 3. Variation of molar conductivity with dilution

electrolytes. The graph shows that there are two extremes of behaviour. Firstly, there are electrolytes, such as sodium and potassium chlorides, which exhibit fairly high values of molar conductivity over the whole concentration range considered. Such electrolytes are called *strong electrolytes*. Others show the behaviour exemplified by acetic acid. The molar conductivity remains low until very high dilutions are reached when it increases enormously. Such electrolytes are called *weak electrolytes*, and this class consists mostly of organic acids and bases. It is not always possible to draw a sharp distinction between strong and weak electrolytes, since the behaviour of some solutions, such as zinc sulphate, lies somewhere between the extremes. These are sometimes classed as *intermediate electrolytes*, of which some salts of the transition metals are the most important.

THEORIES OF IONISATION

Although two or three attempts were made during the nineteenth century to account for the origins of the ions in electrolyte solutions, it was not until 1887, when Arrhenius proposed his theory of ionisation, that any degree of success was achieved.

Arrhenius suggested that in a solution of an electrolyte, ions

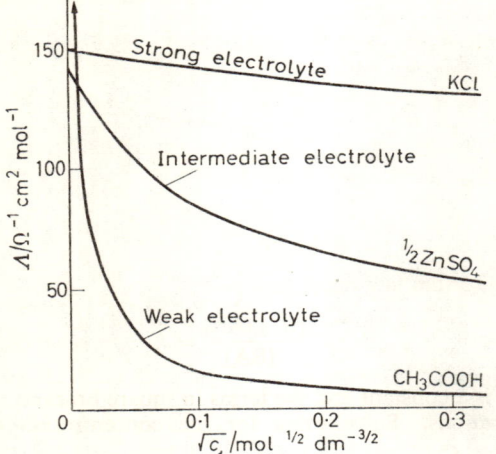

Figure 4. Variation of molar conductivity with $\sqrt{c}$

are always present in equilibrium with un-ionised molecules. Thus for a binary electrolyte, BA, there exists the equilibrium

$$BA \rightleftharpoons B^+ + A^-$$

As the dilution increases, the equilibrium is displaced towards the right and more ionisation occurs until, at infinite dilution, ionisation is complete. Λ^∞, which is the limiting molar conductivity at infinite dilution should be a measure of the total number of ions which can be produced and Λ, at any other concentration

c, should be a measure of the number of ions present at that particular concentration. If the speeds of the ions do not change with concentration, then the ratio of Λ to Λ^{∞} will be equal to the degree of ionisation

$$\frac{\Lambda}{\Lambda^{\infty}} = \alpha \qquad (2.9)$$

where α is the fraction of the solute molecules which have ionised.

Ostwald (1888) applied the equilibrium law to the ionisation of an electrolyte. Consider a binary electrolyte, BA, providing the ions B^+ and A^- in solution where the concentration of BA is c. Suppose a fraction α of the BA ionises so that the concentration of un-ionised BA is $c(1 - \alpha)$ and the concentration of B^+ and A^- ions is αc for each ion

$$\begin{array}{ccc} BA & \rightleftharpoons B^+ & + A^- \\ c(1-\alpha) & \alpha c & \alpha c \end{array}$$

The equilibrium law states that

$$\frac{[B^+][A^-]}{[BA]} = k \qquad (2.10)$$

where k is a constant and the terms in square brackets represent concentrations. Substituting for the concentration terms in (2.10)

$$\frac{\alpha c \cdot \alpha c}{c(1-\alpha)} = k$$

or

$$\frac{\alpha^2 c}{(1-\alpha)} = k \qquad (2.11)$$

This relationship is known as *Ostwald's dilution law*.

If the values of α derived from conductance measurements are correct, then substitution in the left-hand side of eqn. (2.11) should give a constant. Values for acetic acid are given in *Table 3*. Similar results are obtained for other weak electrolytes and it would appear that the Arrhenius theory is satisfactory for this group. Data for strong electrolytes, such as that given for potassium chloride in *Table 4*, show that the theory fails in their case. The only conclusion to be drawn is that, for strong electrolytes, Λ/Λ^{∞} is not equal to the degree of ionisation.

Table 3
DATA FOR ACETIC ACID AT 25°C; $\Lambda^{\infty} = 390 \cdot 7 \, \Omega^{-1} \, cm^2 \, mol^{-1}$

$\dfrac{c}{mol \, dm^{-3}}$	$\dfrac{\Lambda}{\Omega^{-1} \, cm^2 \, mol^{-1}}$	α	$\dfrac{10^5 \alpha^2 c/(1-\alpha)}{mol \, dm^{-3}}$
0·001	48·63	0·1245	1·77
0·005	22·80	0·05835	1·81
0·01	16·20	0·04150	1·80
0·02	11·57	0·02963	1·81
0·05	7·36	0·01884	1·81
0·10	5·20	0·01331	1·80

Table 4
DATA FOR POTASSIUM CHLORIDE AT 25°C; $\Lambda^{\infty} = 149 \cdot 86 \, \Omega^{-1} \, cm^2 \, mol^{-1}$

$\dfrac{c}{mol \, dm^{-3}}$	$\dfrac{\Lambda}{\Omega^{-1} \, cm^2 \, mol^{-1}}$	α	$\dfrac{\alpha^2 c/(1-\alpha)}{mol \, dm^{-3}}$
0·001	147·0	0·981	0·0506
0·005	143·6	0·958	0·1093
0·01	141·3	0·943	0·1561
0·02	138·3	0·923	0·2214
0·05	133·4	0·890	0·3600
0·10	129·0	0·861	0·5330

Tables 3 and *4*: after B. E. CONWAY *Electrochemical Data*, Amsterdam, Elsevier, 1952)

With subsequent work, which developed atomic theory and a knowledge of crystal structure, it was found that many of those substances which were strong electrolytes were completely ionised even in the solid state. For example, the crystal of sodium chloride consists of sodium ions and chloride ions held together by the electrostatic forces between them. When the crystal is dissolved in water, the ions are separated from one another and dissociation is complete at all concentrations. It is interesting to inquire why the ions are separated so easily when the sodium chloride crystal is added to water. The bonds holding an ionic crystal together are very strong, as is evidenced by the high melting points of such crystals (e.g. NaCl, m.p. 801°C). The answer lies in Coulomb's inverse square law governing the attractive force between oppositely charged particles. The force between two charges, q_1 and q_2, is proportional to the product of the charges and inversely proportional to the square of the distance, r, between them

$$F \propto \frac{q_1 q_2}{r^2}$$

To convert this relationship into an equality, a proportionality constant, ϵ, is introduced together with a factor of 4π to rationalise the equation

$$F = \frac{q_1 q_2}{4\pi\epsilon r^2}$$

This factor ϵ is a constant for a given medium and is called the permittivity. The value of ϵ for air is only about 1/80 of that for water. In water, therefore, the force between the ions is much less and consequently a much smaller amount of work is necessary to separate them. When the crystal is placed in water, the ions become hydrated and the system loses some free energy, enough to perform the work of separating the ions.

The Arrhenius theory of ionisation will thus not account for the variation of molar conductivity with concentration which is observed with strong electrolytes.

The work which was to lead to an explanation of the behaviour of strong electrolytes was started by Milner (1912). He calculated the distribution of the ions in a solution of a completely dissociated strong electrolyte. A better attempt was made by Debye

23

and Hückel (1923), and the Debye–Hückel theory is the basis of modern views of strong electrolytes.

Consider the conductance of a strong electrolyte which is completely dissociated in solution, for example, sodium chloride. The sodium ions and chloride ions will not be distributed randomly throughout the solvent. Each sodium ion, being positively charged, will attract towards itself the negatively charged chloride ions and tend to repel other sodium ions. These electrostatic forces will be offset to a large extent by the thermal motion of the ions, but on balance every sodium ion will be surrounded by an ion cloud containing more chloride than sodium ions. Conversely, each chloride ion will have its ion cloud containing more sodium than chloride ions.

Suppose now that a current is passed, and let us fix our attention on one particular sodium ion. When the current flows, the sodium ion will move towards the cathode and its ion cloud will move in the opposite direction. The original ion cloud will break up and a new one will form. In practice, a short time is required for the completion of this operation, and this is known as the *time of relaxation*. Thus, before the original ion cloud has decayed, the sodium ion will be off centre and there will be a net backward attraction. The speed of the sodium ion will thus be reduced. This effect is called the *relaxation effect* or the *asymmetry effect*.

Since the ion cloud will contain water molecules by virtue of the hydration of the chloride ions, the sodium ion is subjected to an increased viscous drag owing to the solvent molecules moving in the opposite direction. This provides a further retarding force on the sodium ion, slowing it down still more. This effect is known as the *electrophoretic effect*.

As the solution is diluted, the ions are spread out farther apart and the density of the ion cloud decreases. Hence the interionic attraction forces are less and the speed of the ion increases. Consequently, charge is being transferred at a greater rate, thus giving rise to a greater current, and the molar conductivity of the solution increases. This change will continue until, at infinite dilution, the ions are an infinite distance apart and interionic effects are zero. Under these conditions, i.e. at infinite dilution, the molar conductivity will be a maximum.

On the basis of these arguments Debye and Hückel were able to derive an expression relating the observed molar conductivity, Λ, at a particular concentration c to that at infinite dilution, Λ^∞. The calculations were subsequently modified by Onsager and the result is known as the Debye–Hückel–Onsager equation or, briefly, as the *Onsager equation*

$$\Lambda = \Lambda^\infty - (A + B\Lambda^\infty)\sqrt{c} \qquad (2.12)$$

where A and B are theoretically deducible constants depending on temperature and nature of the solvent. This equation holds fairly accurately for solutions of concentrations up to 10^{-3} mol dm^{-3} and is accurate to within a few per cent for concentrations up to 10^{-2} mol dm^{-3} for uni-univalent electrolytes. The Onager equation thus provides the theoretical justifications for Kohlrausch's empirical relationship.

For concentrated solutions, and even for dilute ones of some multivalent electrolytes, however, the Onsager equation breaks down. One of the reasons for this breakdown was originally pointed out by Bjerrum who showed that, as the concentration of ions increased, pairs of oppositely charged ions could occur in the solution. These ion pairs are formed when oppositely charged ions approach each other sufficiently closely to be considered more or less as a single entity. The net charge on the single entity is zero when two singly charged ions are involved and thus an ion pair will contribute nothing to the conductance of the solution. It must be pointed out, however, that these ion pairs are not stable and continually exchange partners. Ion association in strong electrolytes has been examined mostly by Davies and by Fuoss and Kraus. These workers have postulated equations which account for the conductance of strong electrolytes up to concentrations of 10^{-1} mol dm^{-3}.

It may be thought in view of the Debye–Hückel theory that interionic attraction would play some part in determining the conductance of weak electrolytes. In dilute solutions of weak electrolytes, however, the number of ions present is very small and interionic effects are negligible. With very concentrated solutions, interionic attraction becomes more important.

We have seen already that the ratio Λ/Λ^∞ can have no significance as a degree of ionisation in the case of strong electrolytes, and it will be remembered that this ratio was put equal to the degree of ionisation on the assumption that the speeds of the ions were constant. We know now that the speeds of the ions do change with concentration, and the variation in the ratio Λ/Λ^∞ is really a measure of the variation of ionic speed with concentration in the case of strong electrolytes. For this reason it is preferable to call the ratio Λ/Λ^∞ the *conductance ratio*. Only in the case of dilute solutions of weak electrolytes may it be considered as a degree of ionisation.

LIMITING MOLAR CONDUCTIVITIES OF IONS

The Onsager equation (2.12) or Kohlrausch's empirical equation (2.8) provides a convenient method of determining Λ^∞ for strong electrolytes. If the results of these determinations for various pairs of electrolytes are considered, some interesting facts emerge. In *Table 5*, Λ^∞ values and the differences between them are shown for some pairs of electrolytes.

Table 5

CONDUCTANCE DATA FOR VARIOUS ELECTROLYTES AT $25°C$

Electrolyte pairs	Λ^∞ $\overline{\Omega^{-1}cm^2}$ mol^{-1}	Difference $\overline{\Omega^{-1}cm^2}$ mol^{-1}	Electrolyte pairs	Λ^∞ $\overline{\Omega^{-1}cm^2}$ mol^{-1}	Difference $\overline{\Omega^{-1}cm^2}$ mol^{-1}
KCl NaCl	149·86 126·45	23·41	KCl KNO_3	149·86 144·96	4·90
KNO_3 $NaNO_3$	144·96 121·55	23·41	NaCl $NaNO_3$	126·45 121·55	4·90
KI NaI	150·32 126·91	23·41	$\frac{1}{2}BaCl_2$ $\frac{1}{2}Ba(NO_3)_2$	139·94 135·04	4·90

(After B. E. Conway *Electrochemical Data*, Amsterdam, Elsevier, 1952)

ELECTROLYTIC CONDUCTION

It will be noticed that the difference between the Λ^∞ values for the pairs of salts in the first column of *Table 5* are constant. The only difference between them is in the cation, as each pair has a common anion. The differences in Λ^∞ values in the fourth column can similarly be attributed to those in the anions. These observations led Kohlrausch to postulate his *Law of independent ionic migration*: 'Each ion contributes a definite amount to the total limiting molar conductivity of the electrolyte irrespective of the nature of the other ion.' This law may be stated in the form

$$\Lambda^\infty = \Lambda_+{}^\infty + \Lambda_-{}^\infty \qquad (2.13)$$

where $\Lambda_+{}^\infty$ and $\Lambda_-{}^\infty$ are the limiting molar conductivities of the cation and anion, respectively. Experiments which will be described later can be used to determine the fraction of the total current carried by a particular ion in an electrolysis, and from these data the limiting molar conductivities of individual ions may be computed. Some values are given in *Table 6*.

Table 6

LIMITING MOLAR CONDUCTIVITIES AT 25°C

Cation	$\dfrac{\Lambda_+{}^\infty}{\Omega^{-1}cm^2mol^{-1}}$	Anion	$\dfrac{\Lambda_-{}^\infty}{\Omega^{-1}cm^2mol^{-1}}$
H^+	349·8	OH^-	197·6
Na^+	50·11	Cl^-	76·34
K^+	73·52	$NO_3{}^-$	71·44
Ag^+	61·92	$\frac{1}{2}SO_4{}^{2-}$	80
$\frac{1}{2}Mg^{2+}$	53·06	CH_3COO^-	40·9
$\frac{1}{2}Ca^{2+}$	59·50	I^-	76·8

(After B. E. CONWAY *Electrochemical Data*, Amsterdam, Elsevier, 1952)

The molar conductivity of an ion is a measure of the amount of current it can carry. As current is the rate of transfer of electric charge, a comparison of the molar conductivities of various ions has more significance if the molar conductivities used all refer to the amounts of ions carrying the same charge.

27

It is thus more meaningful to compare $\Lambda(Na^+)$ with $\Lambda(\frac{1}{2}Mg^{2+})$ rather than with $\Lambda(Mg^{2+})$. In this way the molar conductivities of equivalent amounts of charge carriers are compared.

From *Table 6* it can be seen that

$$\Lambda^\infty(\tfrac{1}{2}Mg^{2+}) = 53 \cdot 06\ \Omega^{-1}\ cm^2\ mol^{-1}$$

If, however, a mole of magnesium ion refers to the entity Mg^{2+}, then

$$\Lambda^\infty(Mg^{2+}) = 106 \cdot 12\ \Omega^{-1}\ cm^2\ mol^{-1}$$

It will thus be appreciated that the following types of relationship apply.

$$\Lambda^\infty(\tfrac{1}{2}Mg^{2+}) = \tfrac{1}{2}\Lambda^\infty(Mg^{2+})$$

$$\Lambda^\infty(2Cl^-) = 2\Lambda^\infty(Cl^-)$$

It has been mentioned above that it is better to compare the molar conductivities of amounts of ions which carry the same amount of charge. The simplest case is obviously the one where the amounts of ions chosen carry unit charge. In order to arrive at the molar conductivity of such an amount of ion the simplest procedure is to divide the molar conductivity for any given amount of ion by the charge number of that amount of ion which quantity may be expressed as Λ_i/z_i. This quantity may be termed *equivalent conductivity* but it should be noted that as the charge number of an ion is dimensionless the equivalent conductivity as defined above will have the same dimensions as molar conductivity. One or two examples should illustrate this point.

$$\Lambda^\infty(Ag^+) = 61 \cdot 92\ \Omega^{-1}\ cm^2\ mol^{-1} \qquad z = 1$$

$$\Lambda^\infty(Ag^+)/z = 61 \cdot 92\ \Omega^{-1}\ cm^2\ mol^{-1}$$

$$\Lambda^\infty(SO_4{}^{2-}) = 160\ \Omega^{-1}\ cm^2\ mol^{-1} \qquad z = 2$$

$$\Lambda^\infty(SO_4{}^{2-})/z = 80\ \Omega^{-1}\ cm^2\ mol^{-1}$$

A further point connected with this matter is that care must be taken in applying the law of independent migration of ions as stated in eqn. (2.13). The molar ionic conductivities should refer to the amounts of the ions contained in the amount of electrolyte specified. Thus, in applying eqn. (2.13) to magnesium chloride, for example, we may write

$$\Lambda^{\infty}(\tfrac{1}{2}MgCl_2) = \Lambda^{\infty}(\tfrac{1}{2}Mg^{2+}) + \Lambda^{\infty}(Cl^-)$$

whence

$$\Lambda^{\infty}(\tfrac{1}{2}MgCl_2) = 53{\cdot}06 + 76{\cdot}34$$
$$= 129{\cdot}40 \; \Omega^{-1} \; cm^2 \; mol^{-1}$$

Alternatively, we may write

$$\Lambda^{\infty}(MgCl_2) = \Lambda^{\infty}(Mg^{2+}) + \Lambda^{\infty}(2Cl^-)$$
$$= \Lambda^{\infty}(Mg^{2+}) + 2\Lambda^{\infty}(Cl^-) \qquad (2.14)$$
$$= 106{\cdot}12 + 152{\cdot}68$$
$$= 258{\cdot}80 \; \Omega^{-1} \; cm^2 \; mol^{-1}$$

It is preferable to state the law of independent migration of ions in a more general form than that of eqn. (2.13) so that the necessity for defining a mole of ion as that amount contained in a mole of electrolyte is avoided.

Suppose that a mole of electrolyte provides in solution v_+ moles of cation and v_- moles of anion,

$$(electrolyte) = v_+(cation) + v_-(anion) \qquad (2.15)$$

Comparison of eqns (2.15) and (2.14) shows that the more general form of the law of independent migration of ions is

$$\Lambda^{\infty} = v_+\Lambda_+{}^{\infty} + v_-\Lambda_-{}^{\infty} \qquad (2.16)$$

One of the great uses of the concept of the independent migration of ions is that it provides a method of calculating Λ^{∞} for weak electrolytes. These values cannot be determined by extrapolation of a plot of Λ^{∞} against $\sqrt{c}$ as with strong electrolytes. Thus to compute for acetic acid we may write

$$\Lambda^{\infty}(CH_3COOH) = \Lambda^{\infty}(H^+) + \Lambda^{\infty}(CH_3COO^-)$$
$$= 349{\cdot}8 + 40{\cdot}9 \; \Omega^{-1} \; cm^2 \; mol^{-1}$$
$$= 390{\cdot}7 \; \Omega^{-1} \; cm^2 \; mol^{-1}$$

This point will be further discussed in Chapter 5.

IONIC MOBILITY AND IONIC CONDUCTANCE

The conductance of an electrolyte is a measure of the current which it can carry, and current is the rate of transfer of electric charge. Charge is carried through the electrolyte by the ions, and hence the conductance of an electrolyte depends on the rate at which the ions can carry charge.

This rate depends on three factors:

(*a*) the number of charges which each ion carries; obviously a polyvalent ion carries more charge with it than a univalent ion.

(*b*) the concentration of the ions; the more ions there are present, the greater the rate at which charge may be transferred.

(*c*) the speed of the ions; once again, the faster an ion travels the greater the rate of transfer of charge.

The speed of the ion, however, in turn depends upon four factors:

(*i*) Electric field intensity—Current flows through the solution by virtue of charge being transferred between the electrodes. A difference of potential is impressed across the electrodes, and this gives rise to an electric field in the solution. A charged particle in an electric field experiences a force which is proportional to the electric field intensity.

(*ii*) The viscosity of the solvent—If the only force on the ion were that provided by the electric field, the ion would accelerate continuously. The viscous drag of the solvent, increasing with velocity, opposes the accelerating force. After a very short time, then, the ion is travelling with a uniform velocity.

(*iii*) The asymmetry effect—This has already been discussed. The effect retards the ion; its extent depends upon concentration.

(*iv*) The electrophoretic effect—This factor has also been discussed already and has been seen to retard the ion by increasing the effective viscous drag of the solvent; its extent likewise depends upon concentration.

It is now necessary to consider the conductance of an electrolyte in terms of the speeds of the ions.

Consider a solution of an electrolyte of concentration c. Suppose that the degree of ionisation is α. (For strong electrolytes, $\alpha = 1$ but the term is retained so that the argument is equally applicable to strong and weak electrolytes.) Let the concentrations of the cations and anions be c_+ and c_- respectively.

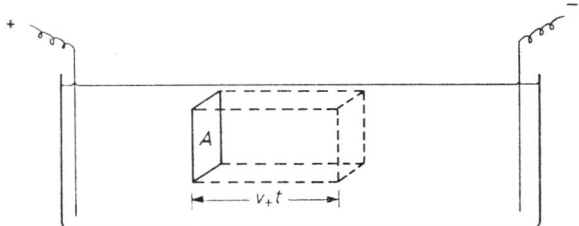

Figure 5. Electrolytic conduction

Let the intensity of the externally applied electric field be E and let the velocities of the cation and anion in this field, at the given concentration, be v_+ and v_- respectively.

Consider a plane of area A, located in the solution perpendicular to the direction of flow of the current as shown in *Figure 5*. In a time t, any cations which were originally located in the plane A will have travelled a distance v_+t towards the cathode. The amount of charge transferred across the area A in a time t by the cations will thus be equal to the amount of charge carried by the cations which are confined within a rectangular block of cross-sectional area A and length v_+t. The volume of this block of solution is v_+tA and the concentration of cations in the solution is c_+. The amount of cation contained in the block is thus c_+v_+tA. The charge carried by a mole of cations is the product of the charge number, z_+, and Faraday's constant, F. The amount of charge transferred across the area A by the cations in a time t is $c_+v_+tAz_+F$.

Similarly, the anions will carry an amount of charge $c_-v_-tA|z_-|F$ across the area A in a time t. In this case we use the modulus of the charge number $|z_-|$ which means the magnitude of z_- but neglecting its sign. This is because we are interested in the total amount of electricity transferred and not in whether it is carried as positive charge or negative charge.

The rate of transfer of charge by the cations and anions respectively is obtained by dividing the quantities $c_+v_+tAz_+F$

31

and $c_- v_- t A |z_-| F$ by the time t. Rate of transfer of charge is equal to current and denoting that carried by the cation across the area A by I_+ and that carried by the anions as I_-, we have

$$I_+ = c_+ v_+ z_+ FA \quad \text{and} \quad I_- = c_- v_- |z_-| FA \quad (2.17)$$

The total current, $I = I_+ + I_-$, hence,

$$I = c_+ v_+ z_+ FA + c_- v_- |z_-| FA \quad (2.18)$$

If a mole of electrolyte provides $_+ v$ cations and v_- anions then,

$$c_+ = v_+ \alpha c \quad \text{and} \quad c_- = v_- \alpha c$$

Substituting in eqn. (2·18)

$$I = v_+ \alpha c v_+ z_+ FA + v_- \alpha c v_- |z_-| FA$$

$$= \alpha c FA (v_+ v_+ z_+ + v_- v_- |z_-|)$$

The current density $j = I/A$, and so

$$j = \alpha c F (v_+ v_+ z_+ + v_- v_- |z_-|) \quad (2.19)$$

As electrolytes obey Ohm's law, the speeds of the ions must be proportional to the electric field intensity, so that

$$v_+ = u_+ E \quad \text{and} \quad v_- = u_- E \quad (2.20)$$

where u_+ and u_- are the proportionality constants and the speeds of the cation and anion, respectively, in unit electric field at the concentration of electrolyte under consideration. The terms u_+ and u_- are called the *mobilities* of the cation and anion respectively, but it must be remembered that, although they refer to unit electric field intensity, they will vary with concentration.

Substituting from eqns (2.20) into (2.19)

$$j = \alpha c F E (v_+ u_+ z_+ + v_- u_- |z_-|)$$

According to eqn. (2.4), the conductivity $\kappa = j/E$, hence

$$\kappa = \alpha c F (v_+ u_+ z_+ + v_- u_- |z|)$$

Furthermore, from eqn. (2.8), $\Lambda = \kappa/c$, hence,

$$\Lambda = \alpha F (v_+ u_+ z_+ + v_- u_- |z_-|) \quad (2.21)$$

At infinite dilution, α will be unity even for weak electrolytes, and if we denote the limiting mobilities of the cation and anion under these conditions by u_+^∞ and u_-^∞ respectively, we may write

$$\Lambda^\infty = F(v_+ z_+ u_+^\infty + v_- |z_-| u_-^\infty) \tag{2.22}$$

Comparing this result with eqn. (2.16) we see that

$$\Lambda_+^\infty = z_+ u_+^\infty F \quad \text{and} \quad \Lambda_-^\infty = |z_-| u_-^\infty F \tag{2.23}$$

Examination of eqn. (2.23) reveals the interesting fact that ions migrate at extremely low speeds during electrolysis. If we consider a univalent cation ($z_+ = 1$) with a limiting molar conductivity of $60\,\Omega^{-1}\,\text{cm}^2\,\text{mol}^{-1}$, and remember that $F = 96\,500\,\text{C mol}^{-1}$

$$u_+^\infty = \frac{\Lambda_+^\infty}{z_+ F} = \frac{60\,\Omega^{-1}\,\text{cm}^2\,\text{mol}^{-1}}{96\,500\,\text{C mol}^{-1}}$$

$$= 6 \cdot 22 \times 10^{-4} \frac{\Omega^{-1}\,\text{cm}^2}{\text{C}}$$

$$= 6 \cdot 22 \times 10^{-4} \frac{\Omega^{-1}\,\text{cm}^2}{\text{A s}}$$

$$= 6 \cdot 22 \times 10^{-4} \frac{\text{cm}^2}{\text{V s}}$$

$$= 6 \cdot 22 \times 10^{-4} \frac{\text{cm s}^{-1}}{\text{V cm}^{-1}}$$

In other words, at infinite dilution the cation which we have considered will move with a speed of $6 \cdot 22 \times 10^{-4}\,\text{cm s}^{-1}$ in an electric field of $1\,\text{V cm}^{-1}$.

In *basic* units the above mobility is given by

$$u_+^\infty = 6 \cdot 22 \times 10^{-8} \frac{\text{m s}^{-1}}{\text{V m}^{-1}}$$

An alternative statement is that the cation moves with a speed of $6 \cdot 22 \times 10^{-8}\,\text{m s}^{-1}$ in an electric field of $1\,\text{V m}^{-1}$, at infinite dilution.

33

TRANSPORT NUMBERS

In electrolysis, all the ions in solution share in carrying the current, and the fraction of the total current carried by a particular species of ion is called the *transport number* or, sometimes, the *transference number* of that ion. The sum of the transport numbers of all the species of ions in solution will thus be equal to unity.

If we consider the situation described in the last section with respect to *Figure 5*, the currents carried by the cations and anions respectively are given by eqn. (2.17)

$$I_+ = c_+ v_+ z_+ FA \quad \text{and} \quad I_- = c_- v_- |z_-| FA$$

The total current is given by

$$I = c_+ v_+ z_+ FA + c_- v_- |z_-| FA$$

and the transport number of the cation, t_+, at electrolyte concentration c, by

$$t_+ = \frac{c_+ v_+ z_+ FA}{c_+ v_+ z_+ FA + c_- v_- |z_-| FA}$$

or

$$t_+ = \frac{c_+ v_+ z_+}{c_+ v_+ z_+ + c_- v_- |z_-|} \tag{2.24}$$

If we consider a uni-univalent electrolyte, then $c_+ = c_-$ and $z_+ = |z_-| = 1$ and eqn. (2.24) reduces to

$$t_+ = \frac{v_+}{v_+ + v_-} \tag{2.25}$$

and similarly for the anion

$$t_- = \frac{v_-}{v_+ + v_-} \tag{2.26}$$

It should be noted that even in this simplest case where a concentration term does not appear explicitly, transport numbers will vary with concentration because they depend on the speeds of the ions which, in turn, depend upon concentration.

As the speeds of the ions are related to their mobilities [cf. eqn. (2.20)], eqn. (2.24) may be equally well expressed

$$t_+ = \frac{c_+ u_+ z_+}{c_+ u_+ z_+ + c_- u_- |z_-|} \tag{2.27}$$

Remembering that $c_+ = v_+ \alpha c$ and $c_- = v_- \alpha c$

$$t_+ = \frac{\alpha c v_+ u_+ z_+}{\alpha c v_+ u_+ z_+ + \alpha c v_- u_- |z_-|}$$

or

$$t_+ = \frac{v_+ z_+ u_+}{v_+ z_+ u_+ + v_- |z_-| u_-} \tag{2.28}$$

Considering the relationship between molar conductivities and mobilities given by eqn. (2.23), for the special case of infinite dilution, eqn. (2.28) could be written

$$t_+^\infty = \frac{v_+ \Lambda_+^\infty}{v_+ \Lambda_+^\infty + v_- \Lambda_-^\infty}$$

or, applying eqn. (2.16)

$$t_+^\infty = \frac{v_+ \Lambda_+^\infty}{\Lambda^\infty} \tag{2.29}$$

Similarly for the anion

$$t_-^\infty = \frac{v_- \Lambda_-^\infty}{\Lambda^\infty} \tag{2.30}$$

From these two relationships, limiting ionic molar conductivities may be calculated, having determined Λ^∞ and t^∞ experimentally. The latter is usually found by extrapolation of a series of measurements over a low concentration range.

It should be noted that, from all the above relationships for transport numbers,

$$t_+ + t_- = 1 \tag{2.31}$$

It must be remembered that a transport number by itself has no meaning. Transport numbers depend upon the speeds of the ions which, in turn, depend upon concentration and temperature, the latter affecting the viscosity of the solvent. Moreover, any other ions present will have some effect on the share of the total current carried by a particular species of ion. A transport number must therefore always be accompanied by a statement of the concentration, temperature and, most important, the electrolyte.

DETERMINATION OF TRANSPORT NUMBERS

Hittorf's method—During electrolysis, the concentration of electrolyte in the anolyte and the catholyte changes as a result of the reactions at the electrodes and the migration of the ions through the solution. The transport numbers of the ion may be calculated from these concentration changes.

Consider the electrolysis of an electrolyte in the cell shown in *Figure 6*. The cell is divided arbitrarily into three compartments, one for the anode, one for the cathode and a central one. Suppose that the electrodes are inert and that current is passed. For the passage of a quantity of electricity Q, the cation will carry a charge Qt_+ and the anion will carry a charge Qt_-. The amount of charge carried by one mole of an ion of type i will be $|z_i|F$. Hence, a charge Qt_+ is carried by an amount of cation Qt_+/z_+F and a charge Qt_- is carried by an amount of anion $Qt_-/|z_-|F$. Thus for the passage of a charge Q, the amount of cation migrating towards the cathode will be Qt_+/z_+F and the amount of anion migrating towards the cathode will be $Qt_-/|z_-|F$.

Consideration of the middle compartment shows that, for the passage of a charge Q, the amount of cation migrating out of the compartment into the catholyte is Qt_+/z_+F and the same amount of cation migrates into the compartment from the anolyte. Similarly, an amount $Qt_-/|z_-|F$ of anion migrates out of the middle compartment into the anolyte and the same amount of anion migrates into the middle compartment from the catholyte. There is therefore, no net change in the concentration of the electrolyte in the middle compartment.

When considering changes of concentration in the anolyte and catholyte, it will probably be better to examine a specific case. Consider the electrolysis of copper(II) chloride with a copper cathode and a platinum anode. At the cathode, copper will be deposited and at the anode chloride ions will be discharged as chlorine gas. Let us consider the changes in concentration of the catholyte and anolyte which result from these actions.

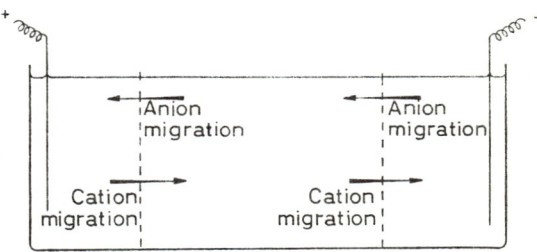

Figure 6. Transport numbers

Catholyte

Loss of Cu^{2+} by discharge $\qquad\qquad \dfrac{Q}{2F} Cu^{2+}$

Gain of Cu^{2+} by migration $= Qt_+/z_+F \qquad\qquad \dfrac{Qt_+}{2F} Cu^{2+}$

Net loss of $Cu^{2+} = (1 - t_+)Q/zF \qquad\qquad \dfrac{Qt_-}{2F} Cu^{2+}$

Loss of Cl^- by migration $= Qt_-/|z_-|F \qquad\qquad \dfrac{Qt_-}{F} Cl^-$

Loss of $CuCl_2$ in catholyte

$\quad = \dfrac{Qt_-}{2F} (Cu^{2+}) + \dfrac{Qt_-}{F} (Cl^-) \qquad\qquad \dfrac{Qt_-}{2F} CuCl_2$

Anolyte

Loss of Cl^- by discharge $\qquad\qquad \dfrac{Q}{F} Cl^-$

Gain of Cl^- by migration $= Qt_-/|z_-|F \qquad\qquad \dfrac{Qt_-}{F} Cl^-$

Net loss of $Cl^- = (1 - t_-) Q/F$ $\qquad\qquad \dfrac{Qt_+}{F} Cl^-$

Loss of Cu^{2+} by migration $= Qt_+/z_+F$ $\qquad \dfrac{Qt_+}{2F} Cu^{2+}$

Loss of $CuCl_2$ in anolyte

$$= \frac{Qt_+}{2F} (Cu^{2+}) + \frac{Qt_+}{F} (Cl^-) \qquad\qquad \frac{Qt_+}{2F} CuCl_2$$

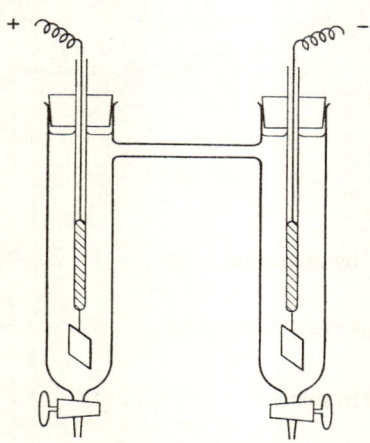

Figure 7. Hittorf cell

The charge passed during an electrolysis may be determined by including a coulometer in the circuit, and if the loss of electrolyte in the anolyte and catholyte is determined by analysis, the transport numbers of the ions may be calculated.

This is the basis of Hittorf's method of determining transport numbers. The experiments are usually carried out in an Hittorf cell, as shown in *Figure 7*.

The anode compartment and the cathode compartment are joined by a thin horizontal tube which minimises diffusion between the anolyte and catholyte. The vertical limbs are equipped with taps to enable samples of the electrode solutions to be run off for analysis.

When determining transport numbers by Hittorf's method it is preferable to avoid systems which involve the evolution of gas at the electrodes, as this may result in stirring and intermixing of the various parts of the solution. Usually evolution of gases is avoided by using, where possible, electrodes consisting of the metal which is the cation in the electrolyte. In these cases, the anode dissolves during electrolysis and metal is plated out on the cathode. Such a system leads to a modified consideration of *Figure 6*.

Consider for example the electrolysis of copper(II) sulphate between copper electrodes.

Catholyte

Loss of Cu^{2+} by discharge $\qquad\qquad\qquad \dfrac{Q}{2F}\,Cu^{2+}$

Gain of Cu^{2+} by migration $= Qt_+/z_+F \qquad \dfrac{Qt_+}{2F}\,Cu^{2+}$

Net loss of $Cu^{2+} = (1-t_+)Q/2F \qquad \dfrac{Qt_-}{2F}\,Cu^{2+}$

Loss of SO_4^{2-} by migration $= Qt_-/|z_-|F \qquad \dfrac{Qt_-}{2F}\,SO_4^{2-}$

Loss of electrolyte in catholyte $\qquad\qquad \dfrac{Qt_-}{2F}\,CuSO_4$

Anolyte

Gain of Cu^{2+} by dissolution of anode $\qquad \dfrac{Q}{2F}\,Cu^{2+}$

Loss of Cu^{2+} by migration $= Qt_+/z_+F \qquad \dfrac{Qt_+}{2F}\,Cu^{2+}$

Net gain of $Cu^{2+} = (1-t_+)\,Q/2F \qquad \dfrac{Qt_-}{2F}\,Cu^{2+}$

Gain of SO_4^{2-} by migration $= Qt_-/|z_-|F$ $\qquad \dfrac{Qt_-}{2F} SO_4^{2-}$

Gain of electrolyte in anolyte $\qquad\qquad\qquad \dfrac{Qt_-}{2F} CuSO_4$

In this case, instead of there being a loss of electrolyte at both electrodes, there is a *gain* of electrolyte at the anode. Furthermore, the change at each electrode is related to the transport number of the anion, and that of the cation cannot be independently determined but must be calculated from the relationship

$$t_+ + t_- = 1$$

If a solution of sodium chloride is to be examined, it is obvious that sodium electrodes cannot be used. With inert electrodes, hydrogen would be evolved at the cathode and chlorine at the anode. The evolution of these gases may still be avoided by using a silver anode and a cathode of silver coated with silver chloride. At the anode the chlorine reacts with the silver to form silver chloride and at the cathode the hydrogen reduces the silver chloride to silver.

Moving boundary methods

We see from eqn. (2.25) and (2.26) that transport numbers are related to the speeds of the ions. Hence, if we can measure these speeds, we can calculate transport numbers. Moving boundary methods of determining transport numbers depend upon this principle which may be illustrated by a consideration of the vertical cell (*Figure 8*).

The cell contains a solution of a uni-univalent electrolyte, *MA* (e.g. KCl) at concentration *c*, together with two other solutions called indicator solutions. One indicator solution, *M'A* (e.g. LiCl), contains a salt with a different cation but the same anion as the solution under study. The other indicator solution, *MA'* (e.g. CH_3COOK), contains a salt with the same cation but a different anion.

The solution *MA* is placed between the indicator solutions in such a way that the most dense solution is at the bottom of the cell. This prevents mixing, and boundaries at *a* and *b* are formed

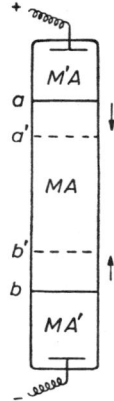

Figure 8. Moving boundary cell

between the solutions. That at *a* is due solely to the difference between the cations *M* and *M′*, as the anions are the same on each side of the boundary. Similarly, the boundary at *b* is due solely to the difference between the anions *A* and *A′*.

On passing current, the cations travel towards the cathode and will carry boundary *a* in this direction also. The anions will travel towards the anode and also carry boundary *b* towards the anode. In order that the boundaries remain sharply defined during the experiment, it is essential that the ion in front of the boundary moves faster than that behind the boundary. If this were not so, the rearmost ion would overtake the foremost, resulting in mixing of the solution, which would lead to the destruction of the boundary. The conditions for sharp boundaries may thus be stated as

Boundary *a*: speed of *M* must exceed speed of *M′*

Boundary *b*: speed of *A* must exceed speed of *A′*

This being the case, the distance which boundary *a* moves in a given time will be proportional to the speed of the cation *M* and

41

the distance which boundary b moves in the same time will be proportional to the speed of the anion A.

Suppose that after a certain time boundary a has moved to a' and boundary b to b'. Then

$$aa' \propto v_+ \quad \text{and} \quad bb' \propto v_-$$

As

$$t_+ = \frac{v_+}{v_+ + v_-} \quad \text{and} \quad t_- = \frac{v_-}{v_+ + v_-}$$

then

$$t_+ = \frac{aa'}{aa' + bb'} \quad \text{and} \quad t_- = \frac{bb'}{aa' + bb'}$$

There are numerous examples of apparatus in which moving boundary experiments may be carried out. The simplest type to use is probably that where the boundary is automatically formed at the beginning of the experiment. Such a boundary is termed an *autogenic boundary*. An autogenic moving boundary apparatus is illustrated in *Figure 9*. It consists of a narrow vertical tube of uniform cross-sectional area, calibrated for volume. An anode of copper or cadmium is attached at the lower end of the tube, while its upper end is connected to a vessel containing a suitable cathode. If potassium chloride solution were so studied, the cathode would probably be of silver coated with silver chloride. The whole apparatus is filled with the solution under study and current is passed. The cadmium anode dissolves to form cadmium chloride which acts as the indicator solution. The boundary between cadmium chloride and potassium chloride is thus generated automatically at the anode surface immediately the current is switched on and continues to move up the tube during the course of the experiment.

Suppose the boundary moves through a volume V in a time t and that during this time the current is kept constant at a value of I. In the time t, then, a quantity of charge It will flow past every point in the circuit. The fraction of this quantity of electricity carried by the cation will be equal to its transport number, t_+. Hence a charge Itt_+ passes every point in the apparatus. As the charge carried by the cation is z_+F, the amount of cation required to carry the charge passed is Itt_+/z_+F. If the

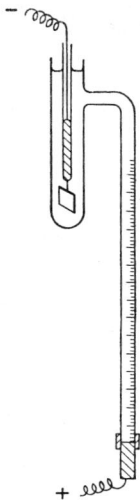

Figure 9. Autogenic moving boundary apparatus

concentration of cation in the solution is c_+, then the cation will sweep through a volume Itt_+/z_+Fc_+ in the time t. Hence

$$V = \frac{Itt_+}{z_+Fc_+}$$

and

$$t_+ = \frac{c_+z_+FV}{It}$$

HYDRATION OF IONS

That most ions in aqueous solution are hydrated has been mentioned earlier, and transport number measurements have leant support to the idea. Such measurements have been used to calculate the limiting molar conductivities of the individual ions by means of eqn. (2.26), and these conductances are a

43

measure of the speed with which the ions travel. The limiting molar conductivities of three of the alkali metal ions (*Table 7*) show an interesting feature.

Table 7

LIMITING MOLAR CONDUCTIVITIES AT 25°C

Ion	Li+	Na+	K+
$\Lambda_+ \infty / \Omega^{-1} \, cm^2 \, mol^{-1}$	38·69	50·11	73·52

The lithium ion is the smallest and lightest of the three ions, the size and weight increasing through sodium to potassium. It would thus be expected that the lithium ion would travel fastest and hence have the highest conductance of the three. Exactly the reverse is true.

To explain this discrepancy Bredig suggested that, in general, ions are hydrated, and when the ion moves it is actually the whole hydrated complex which moves. If the lithium ion were more hydrated than the sodium ion, and this more than the potassium ion, the discrepancy in speeds could be explained by the differences in the sizes of the ion complexes as a whole.

Nernst suggested that, if an inert substance which did not migrate under the influence of an electric field were included in a Hittorf-type cell, then when the conducting ions carried some water with them, the concentration of the inert substance would change. By measuring the change in concentration of the inert substance the amount of water carried by the ions could be calculated.

Washburn conducted such experiments, using sucrose and raffinose as inert substances and measuring their concentrations with a polarimeter. Only relative values of hydration can be measured in this way, but if we assume that the proton is hydrated with one molecule of water, the relative hydrations of the three alkali metal cations which we have considered are as shown in *Table 8*.

It is seen that the lithium ion is hydrated to a much greater extent than the sodium ion which is more hydrated than the potassium ion. These facts could very well account for the order of the conductances of these ions.

Table 8

RELATIVE HYDRATION OF CATIONS

H+	Li+	Na+	K+
1·0	14·0	8·4	5·4

The order of hydration of the ions shown in *Table 8* is not surprising when we consider how hydration might occur. The water molecule is not linear and there exists an angle of about $104\frac{1}{2}$ degrees between the two hydrogen–oxygen bonds, as shown in *Figure 10a*. More electrons are associated with the oxygen atom than with the two hydrogen atoms and the molecule may be considered to act as an electric dipole (*Figure 10b*). Ions in solution can exert an attractive force on one end of the dipole or

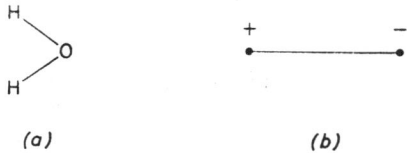

(a)   *(b)*

Figure 10. Water molecule

the other according to whether the ion is a cation or an anion. *Figure 11* illustrates the hydration of a cation by this ion–dipole interaction.

The number of dipoles attracted in this way will depend upon the strength of the electric field at the surface of the ion. This in turn will depend on the radius of the ion. The smaller an ion, the nearer is the central charge to the surface, the greater the electric field and the greater the attraction for the water dipoles. Thus, as lithium is the smallest of the alkali metal ions, we would expect it to suffer the greatest degree of hydration.

This is not the only method of hydration of ions. With some polyvalent ions, the attraction for the water molecules is so great, owing to the increased charge carried by the ions, that the water of hydration is so tightly bound that it may be considered to be coordinated.

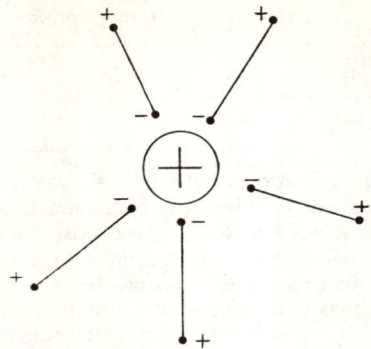

Figure 11. Ion–dipole interaction

TRUE AND APPARENT TRANSPORT NUMBERS

Determination of transport numbers by Hittorf's method depends upon a change of electrolyte concentration, produced by the electrode reactions and the migration of the ions. If we consider the fact that the ions are hydrated and carry some of the solvent with them, we can see that this factor will also affect the concentration changes which take place. If ion hydration is not taken into account when calculating transport numbers from concentration changes, the numbers so obtained are only *apparent*, and a correction must be applied to obtain the *true transport numbers*. Obviously, the greater the concentration of the ions the more solvent is carried about, and the correction which must be applied increases with concentration. In dilute solutions, however, it is so small as to be negligible.

COMPLEX IONS

Sometimes the transport number of an ion turns out to have a negative value or a value greater than unity, as is the case for cadmium iodide. Results for this salt at various concentrations are given in *Table 9*.

Table 9

t_+ AND t_- VALUES FOR CADMIUM IODIDE SOLUTIONS

c/mol dm^{-3}	0·01	0·05	0·2	0·5	1·0	2·0
t_+	0·44	0·36	0·17	0·0	− 0·12	− 0·22
t_-	0·56	0·64	0·83	1·0	1·12	1·22

As the concentration increases, the transport number of the cadmium ion decreases to zero and then becomes negative. Such results may be attributed to the formation of complex ions. Cadmium ions and iodide ions combine to form a complex ion which can probably be represented by $[CdI_4]^{2-}$

$$Cd^{2+} + 4I^- \rightleftharpoons [CdI_4]^{2-}$$

By virtue of the fact that this complex ion is an anion it will travel towards the anode, thereby transporting cadmium to the anode instead of the cathode. The concentration of the complex ion increases with the concentration of the solution, and thus a larger and larger proportion of the cadmium is transferred towards the anode instead of the cathode. At an electrolyte concentration of 0·5 mol dm^{-3}, as much cadmium is travelling towards the anode in the form of the complex ion as is travelling towards the cathode in the form of the simple cation. At higher concentrations still, so much of the complex ion has been formed that more cadmium is transported to the anode than to the cathode, with the result that the transport number of the cadmium ion becomes negative. As the sum of the cation and anion transport numbers must be unity, that of the iodide ion becomes greater than one.

3

IONIC EQUILIBRIA

THE LOWRY–BRØNSTED THEORY OF ACIDS AND BASES

ACIDS and bases were originally defined by Arrhenius as substances which gave rise to hydrogen ions and hydroxide ions, respectively, in solution. Neutralisation was then regarded as the combination of these ions to form un-ionised water

$$H^+ + OH^- \rightarrow H_2O$$

These definitions were perfectly adequate whilst the solvent was water, but when interest in non-aqueous solvents developed, the definitions of acids and bases required modification.

A new definition of acids and bases was put forward by Lowry and Brønsted independently in about 1922. The new definitions do not displace the old ideas but merely extend the concepts of acids and bases and associated phenomena. These modified views have come to be known as the *Lowry–Brønsted theory of acids and bases*.

In this theory, an acid is defined as a substance which will give up a proton, and a base as one which will accept a proton. An acid and a base may thus be related by the equilibrium

$$\underset{acid}{A} \rightleftharpoons H^+ + \underset{base}{B}$$

When an acid and a base differ by one proton, as in the above example, they are said to be *conjugate* to one another. Every acid must have its conjugate base and every base must have its conjugate acid.

When an acid reacts with a base, the products of the reaction are always the conjugate base of the original acid and the conjugate acid of the original base, e.g.

$$A_1 + B_2 \rightleftharpoons A_2 + B_1$$

where A_1 and B_1 are conjugate, as are A_2 and B_2. Such reactions may be regarded simply as proton transfer reactions, as each acid may be considered to consist of its conjugate base plus a proton:

$$\underset{A_1}{(B_1 + H^+)} + B_2 \rightleftharpoons \underset{A_2}{(B_2 + H^+)} + B_1$$

It is unlikely that free protons can exist in solution, so that a potential acid will not be able to function as an acid, when it is the only solute, by giving up a proton unless the solvent will act as a proton acceptor. Conversely, a potential base will not be able to function as a base, when it is the only solute, by accepting a proton unless the solvent will act as a proton donor. Thus the acidic or basic functions of a species cannot become apparent unless the solvent itself is either basic or acidic. This is illustrated by the following examples where water is the solvent.

$$\underset{\substack{acid\ 1}}{CH_3COOH} + \underset{\substack{solvent \\ (base\ 2)}}{H_2O} \rightleftharpoons \underset{\substack{acid\ 2}}{H_3O^+} + \underset{\substack{base\ 1}}{CH_3COO^-}$$

$$\underset{\substack{acid\ 1}}{NH_4^+} + \underset{\substack{solvent \\ (base\ 2)}}{H_2O} \rightleftharpoons \underset{\substack{acid\ 2}}{H_3O^+} + \underset{\substack{base\ 1}}{NH_3}$$

$$\underset{\substack{base\ 1}}{CN^-} + \underset{\substack{solvent \\ (acid\ 2)}}{H_2O} \rightleftharpoons \underset{\substack{acid\ 1}}{HCN} + \underset{\substack{base\ 2}}{OH^-}$$

These examples show that, according to circumstances, water can behave either as an acid or a base. Moreover, the acetate ion is seen to be a base on the Lowry–Brønsted theory, as it is quite capable of accepting a proton to form an acetic acid molecule. This theory thus extends the original Arrhenius theory to cover more substances which may be classed as acids and bases. It is interesting to note that, if we restrict ourselves to water as a solvent, any Lowry–Brønsted acid (e.g. acetic acid or ammonium ions) still produces hydrogen ions (H_3O^+) in solution, and Lowry–Brønsted bases (e.g. cyanide ions) still produce hydroxide ions in solution. In this way the Arrhenius concepts of acid and base (which applied to aqueous solutions only) are not displaced by but absorbed into the wider Lowry–Brønsted ideas.

The above examples also demonstrate the importance of the nature of the solvent as a factor in deciding whether a substance will exhibit acidic or basic properties in solutions. For this reason, solvents are divided into three main classes:

(*a*) *Protogenic solvents* which can give up protons and thus have acidic properties

(*b*) *protophilic solvents* which can accept protons and thus behave as bases

(*c*) *aprotic solvents* which can neither accept nor donate protons and thus preclude any solute from showing acidic or basic functions.

If a solvent, for example water, can be either protogenic or protophilic according to circumstances, it is said to be *amphiprotic*.

INFLUENCE OF THE SOLVENT ON THE STRENGTHS OF ACIDS AND BASES

If we denote an acid as HA and consider it to be dissolved in a protophilic solvent such as water, the equilibrium may be written

$$\underset{acid}{HA} + \underset{solvent}{H_2O} \rightleftharpoons H_3O^+ + A^-$$

The entity H_3O^+ is usually called the hydrogen ion, but it must be remembered that this term really implies ' solvated proton '. It is now believed that the proton is solvated by four water molecules in aqueous solution, $H(H_2O)_4^+$, but the hydrated proton will still be represented as H_3O^+ for convenience. The greater the concentration of hydrogen ions produced by the acid, the stronger is the acid said to be. If the above equilibrium lies far over to the right, the acid is behaving as a strong acid. The position of the equilibrium is determined by two factors: the tendency of the acid to donate a proton and that of the solvent to accept a proton. Water is a very strongly protophilic substance, and when it is used as a solvent, the above reaction occurs to a large extent regardless of whether the acid has a greater or a lesser tendency to donate a proton. It is for this reason that those acids which are normally regarded as strong acids in water (e.g. H_2SO_4, HNO_3, HCl, $HClO_4$) behave as though they were almost equally strong. The differences in the tendencies of these acids to give up protons are dwarfed by the great affinity for protons shown by water.

If a solvent is only weakly protophilic, differences in the strengths of the above acids are immediately apparent. Glacial acetic acid is an example of a weakly protophilic solvent. Acetic acid dissolved in water behaves protogenically, but when used as a solvent in the pure state it will accept protons, if a proton donor with a sufficiently high tendency to yield protons is present. The interaction of an acid HA with glacial acetic acid as solvent may be represented by

$$\underset{acid}{HA} + \underset{solvent}{CH_3COOH} \rightleftharpoons CH_3COOH_2^+ + A^-$$

In this case, the entity $CH_3COOH_2^+$ is the solvated proton and is referred to as the 'hydrogen ion'. The reaction cannot go very far to the right, as the solvent is only weakly protophilic, and the extent of formation of hydrogen ions in this case depends very much upon the proton-donating properties of the solute. Even amongst strong acids there are differences, and the order of decreasing strength is found to be $HClO_4$, HBr, H_2SO_4, HCl, HNO_3. Even perchloric acid is a poor conductor in glacial acetic acid because the degree of ionisation depends on the extent to which the solvent will accept protons.

When the solvent is protogenic itself, with no basic properties, even the strongest acid is incapable of producing hydrogen ions, as there are no molecules available to accept protons. Indeed, many substances which are normally acids, but which are protophilic to some extent, are forced to behave as bases in a strongly protogenic solvent. Nitric acid in hydrogen fluoride is an example of this behaviour:

$$\underset{base}{HNO_3} + \underset{solvent}{HF} \rightleftharpoons H_2NO_3^+ + F^-$$

51

The entity $H_2NO_3^+$ is not a hydrogen ion, as it is not a solvated proton, the solvent being HF.

It is clear then that, for acids as solutes, the more protophilic the solvent, the more readily will the acidic properties of the solute become evident. It will also be appreciated, by an exactly similar series of arguments, that for bases as solutes, the more protogenic the solvent, the more readily will the basic properties of the solute be apparent.

THE CONCEPT OF ACTIVITY

It has already been seen in Chapter 2 that the ions in a solution of an electrolyte cannot be considered as isolated entities. Their behaviour in determining the conductance of an electrolyte is influenced by the presence of neighbouring ions owing to the electrostatic forces which operate. In the same way, the availability of an ion to take part in a reaction or to determine the position of an equilibrium will be affected by the other ions in the solution. It will be appreciated that the concentration of an ion is not a true reflection of its ability to determine any property of the solution, except at infinite dilution where ionic interaction has been eliminated.

It would be better to use some quantity other than concentration to express the availability of an ion to determine the properties of the solution. The quantity which is used is called the *relative activity* of the ion and it takes into account the interactions between the ion and its surroundings. Relative activity may be related to concentration by the equation

$$a = yc \qquad (3.1)$$

where a is the relative activity of the ion and y is called the *activity coefficient*. This is the factor by which the activity differs from the concentration, and it is thus a measure of the interaction between the ions.

It has been shown that ionic interaction varies with concentration and we should thus expect the activity coefficient to vary with concentration. At infinite dilution where there is no ionic interaction the activity coefficient is taken as unity and the relative activity of the ion is equal to its concentration, as we should expect, there being no interaction between ions at infinite dilution. As the concentration increases, the activity coefficient decreases at first, passes through a minimum and then rises, often to values greater than unity in very concentrated solutions. The initial fall is due to increasing interaction between the ions and the subsequent rise is due to non-ionic effects. This variation of the activity coefficient is more marked for polyvalent ions which carry more electric charge and thus are subject to greater electrical forces.

An equation which represents the variation of ionic activity coefficients with concentration has been deduced theoretically by Debye and Hückel, and as the concentration tends to zero it takes the limiting form

$$\log y_i = - A z_i^2 \sqrt{I} \tag{3.2}$$

where y_i is the activity coefficient of an ion of type i, A is a constant, z_i the number of charges carried by an ion of type i and I the ionic strength of the solution. The ionic strength of a solution is given by the relation

$$I = \tfrac{1}{2} \Sigma c_i z_i^2 \tag{3.3}$$

where c_i is the concentration of ions of type i. The summation is carried out over all the types of ions present in the solution.

The constant A of eqn. (3.2) depends upon the temperature and upon the dielectric constant of the solvent. For aqueous solutions at 25°C it has the value $0.51 \ \mathrm{mol}^{-1/2} \ \mathrm{dm}^{3/2}$.

The Debye–Hückel limiting equation (3.2) applies only to infinitely dilute solutions but will give results to within a few per cent for solutions of uni-univalent electrolytes of concentration up to $0.1 \ \mathrm{mol} \ \mathrm{dm}^{-3}$. For more concentrated solutions it has to be modified.

Sometimes when the concentration of ions in a solution is low, with resultantly small ionic interactions, the activity coefficients of the ions are very close to unity and concentrations may be used directly instead of relative activities. The Debye–Hückel limiting equation enables us to calculate the limits of concentration below which we can justifiably use concentrations instead of relative activities.

Suppose we make the assumption that, if the values of the ionic activity coefficients lie between 0·95 and 1·00, we can afford to consider the activities of the ions to be the same as their concentrations. Consider now a 0·1 mol dm^{-3} aqueous solution of acetic acid at 25°C. *Table 3* (p. 22) shows that the degree of dissociation of acetic acid under these conditions is 0·01331, and this means that the concentration of hydrogen ions and acetate ions will be (0·1 × 0·01331) mol dm^{-3}. The ionic strength of the solution is given by

$$I = \tfrac{1}{2} [c_{H^+}z^2_{H^+} + c_{Ac^-}z^2_{Ac^-}]$$

where the symbol H^+ denotes the hydrogen ion and Ac^- denotes the acetate ion.

Hence

$$I = \tfrac{1}{2}[(0·1 × 0·01331 × 1^2) + (0·1 × 0·01331 × 1^2)]$$
$$= 0·001331 \text{ mol dm}^{-3}$$

The activity coefficients of the hydrogen ions or the acetate ions will be given by eqn. (3.2)

$$\log y_{H^+} = -0·51 \, z_{H^+}^2 \sqrt{I}$$

$$= -0·51 × 1^2 × \sqrt{0·001331}$$

$$= -0·0186$$

hence

$$\log y_{H^+} = \bar{1}·9814$$

and

$$y_{H^+} = 0·958$$

Similarly, the activity coefficient of the acetate ion will also have the value 0·958. An acetic acid concentration of 0·1 mol dm^{-3} will thus be the upper limit at which we can tolerate ionic concentrations instead of ionic activities, and the same will be true of other weak electrolytes with strengths similar to that of acetic acid.

If we consider a strong electrolyte such as hydrochloric acid which is completely dissociated in aqueous solution, it is apparent that the concentration of ions will be equal to that of the acid, with a resultantly greater degree of ionic interaction. In this case, it is to be expected that the upper limit of concentration where activities can be taken as equal to concentrations will be much lower than for weak electrolytes. Suppose we consider a 0.001 mol dm^{-3} aqueous solution of hydrochloric acid at 25°C. The concentrations of the hydrogen ions and the chloride ions will be 0.001 mol dm^{-3}, as dissociation is complete. The ionic strength of the solution is given by

$$I = \tfrac{1}{2} [c_{H^+}z^2_{H^+} + c_{Cl^-}z^2_{Cl^-}]$$
$$= \tfrac{1}{2} [(0.001 \times 1^2) + (0.001 \times 1^2)]$$
$$= 0.001 \text{ mol } dm^{-3}$$

The activity coefficient of the hydrogen ions is given by

$$\log y_{H^+} = -0.51 \, z_{H^+}^2 \, \sqrt{I}$$
$$= -0.51 \times 1^2 \times \sqrt{0.001}$$
$$= -0.0161$$

hence

$$\log y_{H^+} = \bar{1}.9839$$

and

$$y_{H^+} = 0.964$$

The activity coefficient of the chloride ions will also be 0.964. As these values lie just within the original limits of 0.95–1.00 which we imposed, the upper limit of concentration for a strong electrolyte at which ionic activities may be taken as equal to ionic concentrations is about 0.001 mol dm^{-3}.

In solutions of weak electrolytes, un-ionised solute molecules will be present, and such molecules will be subject to van der Waals-type intermolecular forces. Strictly, we should use activities for un-ionised particles as well as for ions. The intermolecular forces between neutral particles, however, are much weaker than the electrical forces between ions, and hence the activity coefficients of neutral species do not depart very far from unity.

As a general rule for solutions of weak electrolytes it may be said that, if the ionic activity coefficients may be taken as unity, there is usually even more justification for taking those of the neutral solute molecules as equal to unity also.

As the latter suffer some degree of molecular interaction, so do the neutral solvent molecules. In the case of the solvent, however, the relative activity is usually related to the *mole fraction* of solvent in the solution by the relation

$$a = fx \tag{3.4}$$

where x is the mole fraction of solvent, given by

$$x = \frac{n_1}{n_1 + n_2} \tag{3.5}$$

n_1 being the amount of solvent and n_2 that of solute in the solution. When the system consists of pure solvent, $n_2 = 0$ and $x = 1$. As there will be no interaction of the solvent molecules with solute under these conditions, the activity coefficient has the value unity and the relative activity of the solvent is equal to its mole fraction. Thus the relative activity of pure solvent is unity. When a solution is so dilute that the activity coefficients of the various solute species may be taken as unity, the solution approximates to pure solvent and the activity of the solvent in the system may be taken as unity, with negligible error.

DISSOCIATION CONSTANTS OF ACIDS AND BASES

When an acid HA is dissolved in water, the following equilibrium is established

$$HA + H_2O \rightleftharpoons H_3O^+ + A^-$$

When the equilibrium law is applied, it is more correct to use relative activities instead of concentrations, and we have

$$\frac{a_{H^+} \times a_{A^-}}{a_{HA} \times a_{H_2O}} = \text{constant} \tag{3.6}$$

The symbol H^+ is used to represent the hydrated proton. As we have used relative activities, the constant is a true constant.

If the solution is dilute, we may take the relative activity of water to be unity and write

$$\frac{a_{H^+} \times a_{A^-}}{a_{HA}} = K_a \qquad (3.7)$$

K_a is known as the thermodynamic* dissociation constant of the acid; it is a constant for solutions where the mole fraction of the solvent is very close to unity.

Eqn. (3.7) may be written in the alternative form

$$\frac{[H^+][A^-]}{[HA]} \cdot \frac{y_{H^+} y_{A^-}}{y_{HA}} = K_a \qquad (3.8)$$

where the terms in square brackets represent concentrations. If the solution is sufficiently dilute, the activity coefficients may be taken as unity with only a very small error and we may write

$$\frac{[H^+][A^-]}{[HA]} \approx K_a \qquad (3.9)$$

Frequently the expression on the left-hand side of eqn. (3.9) is put equal to k_a which is called the classical dissociation constant. This is not a true constant but varies slightly with concentration. If we restrict our considerations to dilute solutions, however, we can use eqn. (3.9) without introducing any serious errors.

The classical dissociation constant k_a is identical with the constant of Ostwald's dilution law [eqn. (2.11)], and in view of the approximations we have made in arriving at eqn. (3.9) it is remarkable that the data for acetic acid in *Table 3* show such a constancy in the fourth column. The constancy of the dilution law expression is somewhat fortuitous. As concentration increases, the activity coefficients, which are omitted in the dilution law, decrease, and this should give values of k which decrease with increasing concentration. The values of α, however, which were used to calculate k were obtained from conductance measurements on the assumption that the speeds of

---

* The term thermodynamic is included because activity is a thermodynamic concept. For more detailed information on activities and activity coefficients a textbook of thermodynamics should be consulted.

the ions did not vary with concentration. We now know that this is untrue and that the speeds of the ions decrease with increasing concentration, so that the values of α used were too small. This corrects for the omission of the activity coefficients, so that the constancy of k is due to two self-cancelling errors. Both of these errors are very small for dilute solutions.

A situation similar to that for acids exists for bases. If a base B is dissolved in water, the following equilibrium is established

$$B + H_2O \rightleftharpoons BH^+ + OH^-$$

Applying the equilibrium law

$$\frac{a_{BH^+} \times a_{OH^-}}{a_B \times a_{H_2O}} = \text{constant} \tag{3.10}$$

Table 10

DISSOCIATION CONSTANTS OF ACIDS AND BASES IN WATER AT 25°C

Acid	K_a	Base	K_b
acetic	1.75×10^{-5}	ammonia	1.77×10^{-5}
monochloracetic	1.40×10^{-3}	ethylamine	4.66×10^{-4}
propionic	1.34×10^{-5}	diethylamine	1.00×10^{-2}
formic	1.77×10^{-4}	triethylamine	5.24×10^{-4}
benzoic	6.3×10^{-5}	aniline	3.8×10^{-10}

(After B. E. CONWAY *Electrochemical Data*, Amsterdam, Elsevier, 1952)

Assuming that the solution is sufficiently dilute for the relative activity of the water and the activity coefficients of the various solute species to be taken as unity, eqn. (3.10) reduces to

$$\frac{[BH^+][OH^-]}{[B]} = k_b \approx K_b \tag{3.11}$$

where k_b and K_b are the classical and thermodynamic dissociation constants, respectively, of the base.

These acid and base dissociation constants are useful in indicating the strengths of acids and bases. The greater the strength of an acid or of a base, the greater will be the extent of ionisation and the greater will be the dissociation constant.

The dissociation constants of some acids and bases in water are given in *Table 10*.

It is interesting to note how the solvent affects the strength of an acid. The dissociation constant of benzoic acid in water is 6.3×10^{-5}, whereas in a solvent consisting of 20 per cent methanol and 80 per cent water, the dissociation constant drops to 1.9×10^{-5}.

Acids which are polybasic will have a dissociation constant for each stage of dissociation, and the same is true for polyacidic bases. Phosphoric acid, for example, dissociates in three stages, and a dissociation constant is allocated to each

$$H_3PO_4 + H_2O \rightleftharpoons H_3O^+ + H_2PO_4^- \qquad K_{a_1} = 7.11 \times 10^{-3}$$

$$H_2PO_4^- + H_2O \rightleftharpoons H_3O^+ + HPO_4^{2-} \qquad K_{a_2} = 6.34 \times 10^{-8}$$

$$HPO_4^{2-} + H_2O \rightleftharpoons H_3O^+ + PO_4^{3-} \qquad K_{a_3} = 4.73 \times 10^{-13}$$

It will be noticed that these equilibria show that H_3PO_4 is an acid of which the conjugate base is $H_2PO_4^-$, also that $H_2PO_4^-$ can behave as an acid giving rise to its conjugate base HPO_4^{2-} which, in turn, can behave as an acid producing its conjugate base PO_4^{3-}.

The dissociation constants of acids and bases vary with temperature, so that their values must always be accompanied by a statement of the temperature to which they refer.

UNITS OF DISSOCIATION CONSTANTS

Thermodynamic activities are dimensionless and if we consider the thermodynamic dissociation constant of an acid HA, given by

$$K_a = \frac{a_{H^+} \times a_{A^-}}{a_{HA}}$$

it will be understood that thermodynamic dissociation constants are also dimensionless. The classical dissociation constant, however, is given by

$$k_a = \frac{[H^+][A^-]}{[HA]}$$

and as concentration has dimensions, very often ($mol\ dm^{-3}$), it is clear that the classical dissociation constant will have dimensions, depending upon the number of species involved in the equilibrium.

These remarks apply to any equilibrium constants and should provide a means of identifying them as classical or thermo-dynamic.

SELF-IONISATION OF SOLVENTS

No matter how pure, many solvents have a small electrical conductance. This is attributed to ions arising from the solvent molecules by a process of self-ionisation. Liquid ammonia, liquid hydrogen fluoride and water, for example, exhibit this phenomenon:

$$NH_3 + NH_3 \rightleftharpoons NH_2^- + NH_4^+$$

$$HF + HF \rightleftharpoons H_2F^+ + F^-$$

$$H_2O + H_2O \rightleftharpoons H_3O^+ + OH^-$$

When the self-ionisation of a solvent involves the transfer of a proton, as in the above cases, it is sometimes known as *auto-protolysis*. Considering a general solvent, XH, these equilibria may be represented by

$$XH + XH \rightleftharpoons XH_2^+ + X^-$$

Applying the equilibrium law

$$\frac{a_{XH_2^+} \times a_{X^-}}{a_{XH}^2} = \text{constant}$$

As the amount of ionisation is small, we may take the relative activity of the un-ionised solvent as unity and write

$$a_{XH_2^+} \times a_{X^-} = K_I \qquad (3.12)$$

where K_I is a constant known as the *ionic product* of the solvent. Because the concentration of ions is very low, their relative activities are often taken as equal to their concentrations, and eqn. (3.12) is frequently written

$$[XH_2^+][X^-] = K_I$$

In the special case where water is the solvent, the ionic product is denoted by K_w, so that

$$[H^+][OH^-] = K_w \qquad (3.13)$$

where the symbol H^+ once again represents the hydrated proton.

K_w has been determined from accurate conductance measurements on extremely pure water (known as conductivity water) and has been found to have the value 10^{-14} at 25°C.

RELATIONSHIP BETWEEN THE STRENGTHS OF CONJUGATE ACIDS AND BASES

Consider the interaction of an acid HA with an amphiprotic solvent such as water

$$HA + H_2O \rightleftharpoons H_3O^+ + A^-$$

A^- is the conjugate base of the acid and may itself react with the solvent

$$A^- + H_2O \rightleftharpoons HA + OH^-$$

The dissociation constant of the acid HA is

$$K_a = \frac{[H^+][A^-]}{[HA^-]}$$

and the dissociation constant of the conjugate base A^- is

$$K_b = \frac{[HA^-][OH^-]}{[A^-]}$$

Now

$$K_a K_b = \frac{[H^+][A^-]}{[HA]} \cdot \frac{[HA^-][OH^-]}{[A^-]} = [H^+][OH^-]$$

Hence

$$K_a K_b = K_w \qquad (3.14)$$

This result implies that the strengths of a conjugate acid and base in the same solvent must be in inverse ratio to one another. The greater the strength of an acid, and hence the larger K_a, the smaller must K_b be and hence the weaker the conjugate base for eqn. (3.14) to be true.

Because of this relationship it is sometimes the practice to deal only with acid dissociation constants. Information on the strength of a base can be conveyed by quoting the dissociation constant of its conjugate acid provided the ionic product of the solvent is known.

ACIDITY AND ALKALINITY OF AQUEOUS SOLUTIONS

In a sample of pure water, the hydrogen ions and hydroxide ions arise from the solvent molecules:

$$H_2O + H_2O \rightleftharpoons H_3O^+ + OH^-$$

The concentration of hydrogen ions must be equal to that of hydroxide ions and the water is said to be neutral. As $[H^+] = [OH^-]$, we have from eqn. (3.13)

$$[H^+] = [OH^-] = \sqrt{K_W} = \sqrt{10^{-14}} = 10^{-7} \text{ mol dm}^{-3}$$

In neutral solution at 25°C, then, the concentration of hydrogen ions is 10^{-7} mol dm^{-3}. If the hydrogen ion concentration is greater than 10^{-7} mol dm^{-3}, the solution will be acidic; if it is less than 10^{-7} mol dm^{-3}, alkaline.

A convenient scale of acidity and alkalinity is obtained by the definition of pH as the common logarithm of the reciprocal of the hydrogen ion activity:

$$\text{pH} = \log \frac{1}{a_{H^+}} = -\log a_{H^+}$$

For most purposes it is sufficient to take the hydrogen ion activity as equal to the hydrogen ion concentration and write

$$\text{pH} = -\log [H^+] \qquad (3.15)$$

Hence, if

$$[H^+] = 10^{-7} \text{ mol dm}^{-3}, \text{ pH} = 7 \quad \text{neutral solution}$$
$$[H^+] = 10^{-3} \text{ mol dm}^{-3}, \text{ pH} = 3 \quad \text{acidic solution}$$
$$[H^+] = 10^{-9} \text{ mol dm}^{-3}, \text{ pH} = 9 \quad \text{alkaline solution}$$

The concept of pH can be applied to other ions such as hydroxide ions and also to dissociation constants:

$$pOH = - \log [OH^-]$$

and

$$pK_a = - \log K_a$$

Equation (3.13) can be written in the logarithmic form

$$\log [H^+] + \log [OH^-] = \log K_w$$

or

$$- \log [H^+] - \log [OH^-] = - \log K_w$$

and hence

$$pH + pOH = pK_w \qquad (3.16)$$

where $pK_w = 14$ at 25°C.

In a solution of 0·01 mol dm^{-3} hydrochloric acid which is completely dissociated, the concentration of hydrogen ions will also be 0·01 mol dm^{-3}

$$[H^+] = 0.01 \text{ mol dm}^{-3} = 10^{-2} \text{ mol dm}^{-3} \text{ hence pH} = 2$$

In a solution of 0·01 mol dm^{-3} sodium hydroxide which is completely dissociated, the concentration of hydroxide ions will also be 0·01 mol dm^{-3}

$$[OH^-] = 0.01 \text{ mol dm}^{-3} = 10^{-2} \text{ mol dm}^{-3} \text{ hence pOH} = 2$$

As

$$pH + pOH = 14$$

then

$$pH = 12$$

The pH scale thus provides a system where the acidity of a solution can be measured in terms of small positive numbers. For most practical purposes, the scale can be considered to extend from 0 to 14, but in very strongly acidic solutions the pH may be negative and in very strongly alkaline solutions it may be greater than 14.

For neutral solutions, $[H^+] = \sqrt{K_w}$ and thus neutral pH = $\frac{1}{2}pK_w$. Now K_w is of the nature of a dissociation constant and varies with temperature, so that the neutral pH will also vary with temperature. It must thus be remembered that a solution of pH 7 will be neutral only at 25°C.

NEUTRALISATION AND HYDROLYSIS

The Arrhenius theory described neutralisation as the reaction between equivalent amounts of an acid and a base to produce a salt plus water. In the modern sense, neutralisation is still the reaction between equivalent amounts of an acid and a base, but as we have seen already, the products of such a reaction are the conjugate base of the acid and the conjugate acid of the base, e.g.

$$\underset{\text{acid 1}}{HCl} + \underset{\text{base 2}}{CH_3COO^-} \rightleftharpoons \underset{\text{acid 2}}{CH_3COOH} + \underset{\text{base 1}}{Cl^-}$$

In general, neutralisation can be written

$$\underset{\text{acid}}{HA} + \underset{\text{base}}{B} \rightleftharpoons BH^+ + A^-$$

where BH^+ and A^- can be regarded as the ions of the salt produced by the neutralisation. The extent to which the reaction proceeds depends not only on the strengths of the acid and base but also on the nature of the solvent. If the solvent is amphiprotic like water, it may interact with the products of neutralisation in two ways:

$$BH^+ + H_2O \rightleftharpoons B + H_3O^+ \qquad (a)$$

$$A^- + H_2O \rightleftharpoons HA + OH^- \qquad (b)$$

In reaction (a), the original base is produced and in reaction (b) the original acid. The neutralisation process has thus been reversed. This phenomenon is known as *solvolysis* or, in the specific case where water is the solvent, *hydrolysis*.

Reaction (a) is favoured when the base B is weak. When this is the case, the base has only a small tendency to accept a proton and will give it up again quite readily if the opportunity offers, as in reaction (a). Reaction (b) is favoured when the acid HA is weak. If the original acid is weak, it only has a small tendency to lose a proton and will accept a proton back again by reaction (b).

It should also be noted that reaction (a) leads to the generation of hydrogen ions, thus producing an acidic solution, and reaction (b) generates hydroxide ions, thus producing an alkaline solution. If a strong acid reacts with a strong base, no hydrolysis will occur, and if equivalent amounts of acid and base are used, the solution will be neutral. If equivalent amounts of a strong acid and a weak base react to form a salt, the hydrolysis reaction (a), favoured by a weak base, will occur and the resulting solution will be acidic. If equivalent amounts of a weak acid and a strong base react to form a salt, the hydrolysis reaction (b), favoured by a weak acid, will occur and the solution will be alkaline. When equivalent amounts of a weak acid and a weak base react to form a salt, both hydrolysis reactions will occur and the solution will be acidic or alkaline depending on the relative strengths of the acid and the base.

Consider the hydrolysis of the salt of a weak acid and a strong base. In solution, the ions of the salt BH^+ and A^- will be present. As the original base was strong, only the A^- ion will take part in the hydrolysis reaction

$$A^- + H_2O \rightleftharpoons HA + OH^-$$

Applying the equilibrium law

$$\frac{a_{HA} \times a_{OH^-}}{a_{A^-} \times a_{H_2O}} = \text{constant}$$

Assuming that the relative activity of water may be taken as unity

$$\frac{a_{HA} \times a_{OH^-}}{a_{A^-}} = K_h$$

where K_h is called the *hydrolysis constant* of the salt. If relative activities may be taken as equal to concentrations

$$\frac{[HA]\,[OH^-]}{[A^-]} = K_h$$

The dissociation constant of the original acid is given by

$$K_a = \frac{[H^+] [A^-]}{[HA^-]}$$

and the ionic product of water, $K_w = [H^+] [OH^-]$
Now

$$\frac{K_w}{K_a} = \frac{[H^+] [OH^-] [HA]}{[H^+] [A^-]} = \frac{[HA] [OH^-]}{[A^-]}$$

Hence

$$\frac{K_w}{K_a} = K_h \tag{3.17}$$

Comparison of eqn. (3.17) with (3.14), which may be written in the form

$$\frac{K_w}{K_a} = K_b$$

shows that the hydrolysis constant, K_h, of the salt of the acid HA and the base B is equal to the dissociation constant of the conjugate base of the acid HA. The conjugate base of HA is, of course, the ion A^-, and if we consider this as a base reacting with the solvent to produce hydroxyl ions, we have

$$A^- + H_2O \rightleftharpoons HA + OH^-$$

From this reaction the dissociation constant of the base A^- is seen to be

$$K_b = \frac{[HA] [OH^-]}{[A^-]}$$

which is in fact the same as the hydrolysis constant of the salt.
If we consider, for example, the salt sodium acetate, which is that of a weak acid and a strong base, the hydrolysis reaction will be

$$CH_3COO^- + H_2O \rightleftharpoons CH_3COOH + OH^-$$

and the hydrolysis constant of sodium acetate is given by

$$K_h = \frac{[CH_3COOH]\,[OH^-]}{[CH_3COO^-]} \qquad (3.18)$$

Acetic acid reacts with water to produce its conjugate base, the acetate ion

$$CH_3COOH + H_2O \rightleftharpoons H_3O^+ + CH_3COO^-$$

As the acetate ion is a base, it can react with water

$$CH_3COO^- + H_2O \rightleftharpoons CH_3COOH + OH^-$$

This reaction is exactly the same as the hydrolysis reaction but can now be regarded as the dissociation reaction of the base CH_3COO^-, this constant therefore being

$$K_b = \frac{[CH_3COOH]\,[OH^-]}{[CH_3COO^-]} \qquad (3.19)$$

The constants of eqn. (3.18) and (3.19) may be equally well termed either the *hydrolysis constant of sodium acetate* or the *base dissociation constant of the acetate ion*.

Considering the general hydrolysis reaction once again

$$A^- + H_2O \rightleftharpoons HA + OH^-$$

it may be seen that $[HA] = [OH^-]$, so that the hydrolysis constant may be written

$$K_h = \frac{[OH^-]^2}{[A^-]}$$

If the extent of hydrolysis is small (i.e. if $K_h < 10^{-2}$), the concentration of A^- ions will not be very much less than that of the salt in the solution. Denoting the concentration of the salt by c, we may write

$$[A^-] \approx c$$

and

$$K_h \approx \frac{[OH^-]^2}{c}$$

or

$$[OH^-] = \sqrt{K_h c}$$

67

Now $[H^+] [OH^-] = K_w$, so that

$$[H^+] = \frac{K_w}{[OH^-]} = \frac{K_w}{\sqrt{K_h c}}$$

Moreover, $K_h = K_w/K_a$, and hence

$$[H^+] = \frac{K_w}{\sqrt{K_w c/K_a}} = \sqrt{\frac{K_w K_a}{c}}$$

Taking logs

$$\log [H^+] = \tfrac{1}{2} \log K_w + \tfrac{1}{2} \log K_a - \tfrac{1}{2} \log c$$

or

$$pH = \tfrac{1}{2}pK_w + \tfrac{1}{2}pK_a + \tfrac{1}{2} \log c \qquad (3.20)$$

Now consider the hydrolysis of the salt of a strong acid and a weak base. In solution, the ions of the salt BH^+ and A^- will be present. As the original acid was strong, only the BH^+ ion will hydrolyse

$$BH^+ + H_2O \rightleftharpoons B + H_3O^+$$

and the hydrolysis constant K_h will be given by

$$K_h = \frac{[H^+] [B]}{[BH^+]}$$

The dissociation constant of the original base, B, is

$$K_b = \frac{[BH^+] [OH^-]}{[B]}$$

whence

$$\frac{K_w}{K_b} = \frac{[H^+] [OH^-] [B]}{[BH^+] [OH^-]} = \frac{[H^+] [B]}{[BH^+]}$$

and thus

$$\frac{K_w}{K_b} = K_h \qquad (3.21)$$

Once again the ion BH^+ is the conjugate acid of the base B and the hydrolysis reaction could equally well be regarded as the dissociation of the acid BH^+. Thus, if we take the salt anilinium chloride, $C_6H_5NH_3Cl$, as an example of a salt of a strong acid, HCl, and a weak base, $C_6H_5NH_2$, the constant K_h of eqn. (3.21) may be called the *hydrolysis constant of anilinium chloride*, or it might equally well be denoted by the symbol K_a and called the *acid dissociation constant of the anilinium ion*, $C_6H_5NH_3^+$.

In the hydrolysis reaction it may be seen that $[B] = [H^+]$, and hence the hydrolysis constant may be written

$$K_h = \frac{[H^+]^2}{[BH^+]}$$

If the extent of hydrolysis is small ($K_h < 10^{-2}$), the concentration of BH^+ ions will not be much less than that of the salt, and denoting this by c, we have

$$K_h = \frac{[H^+]^2}{c}$$

or

$$[H^+] = \sqrt{K_h c}$$

Now by eqn. (3.21), $K_h = K_w/K_b$, and hence

$$[H^+] = \sqrt{\frac{K_w c}{K_b}}$$

Taking logs

$$\log [H^+] = \tfrac{1}{2} \log K_w - \tfrac{1}{2} \log K_b + \tfrac{1}{2} \log c$$

or

$$pH = \tfrac{1}{2} pK_w - \tfrac{1}{2} pK_b - \tfrac{1}{2} \log c \quad (3.22)$$

For the salt of a weak acid and a weak base, both types of hydrolysis reaction will occur

$$A^- + H_2O \rightleftharpoons HA + OH^-$$

$$BH^+ + H_2O \rightleftharpoons B + H_3O^+$$

Assuming that the activity of water is unity, the equilibrium constants K_1 and K_2 for the first and second reactions, respectively, are given by

$$K_1 = \frac{[HA][OH^-]}{[A^-]} \quad \text{and} \quad K_2 = \frac{[B][H^+]}{[BH^+]}$$

When the two reactions are added together to arrive at the total hydrolysis, we have

$$A^- + BH^+ + 2H_2O \rightleftharpoons H_3O^+ + OH^- + HA + B$$

The equilibrium constant of this reaction is

$$\frac{[H^+][OH^-][HA][B]}{[A^-][BH^+]} = K_1K_2$$

or

$$K_w \frac{[HA][B]}{[A^-][BH^+]} = K_1K_2$$

As K_w, K_1 and K_2 are constants, the expression

$$\frac{[HA][B]}{[A^-][BH^+]}$$

must also be a constant: it is called the hydrolysis constant of the salt. Thus

$$K_h = \frac{K_1K_2}{K_w}$$

Now K_1 is the dissociation constant of the base A^- which is related to K_a, the dissociation constant of its conjugate acid HA by eqn. (3.14)

$$K_1 = K_w/K_a$$

Also the dissociation constant K_2 of the acid BH^+ is related to the dissociation constant K_b of the conjugate base B

$$K_2 = K_w/K_b$$

Hence

$$K_h = \frac{K_w K_w}{K_w K_a K_b}$$

$$= \frac{K_w}{K_a K_b} \qquad (3.24)$$

The calculation of the pH of a solution of the salt of a weak acid and a weak base is rather complicated, but if some simplifying assumptions are made, a useful approximation can be obtained. Suppose that the concentration of the salt is c and that the original acid and base had similar strengths. In this case, the two hydrolysis reactions will proceed to approximately the same extent and we may say that $[HA] \simeq [B]$. Moreover, if the extent of hydrolysis is small ($K_h < 10^{-2}$), the concentration of BH^+ ions and A^- ions will not be much less than that of the salt. We may thus make the second approximation, $[BH^+] = [A^-] = c$. The hydrolysis constant

$$K_h = \frac{[HA][B]}{[BH^+][A^-]}$$

may thus be written

$$K_h = \frac{[HA]^2}{c^2}$$

and

$$[HA] = c\sqrt{K_h} = c\sqrt{\frac{K_w}{K_a K_b}} \qquad (3.25)$$

The dissociation constant, K_a, of the original acid is

$$K_a = \frac{[H^+][A^-]}{[HA]}$$

thus

$$[HA] = \frac{[H^+][A^-]}{K_a}$$

71

Remembering that $[A^-] = c$, we have

$$[HA] = [H^+] \frac{c}{K_a}$$

and substituting in eqn. (3.25)

$$[H^+] \frac{c}{K_a} = c \sqrt{\frac{K_w}{K_a K_b}}$$

or

$$[H^+] = \sqrt{\frac{K_a K_w}{K_b}}$$

This expression may be written in the form

$$pH = \tfrac{1}{2} pK_w + \tfrac{1}{2} pK_a - \tfrac{1}{2} pK_b \qquad (3.26)$$

Equation (3.26) shows that the pH of the solution is independent of the concentration of the salt, but it must be remembered that this will only be true if the original acid and base have very similar strengths and the extent of hydrolysis is limited.

BUFFER SOLUTIONS

Aqueous solutions of both sodium chloride and ammonium acetate have a pH of about 7. Sodium chloride is the salt of a strong acid and a strong base and no hydrolysis occurs, the solution remaining neutral. Ammonium acetate is the salt of a weak acid and a weak base of similar strengths ($K_a = 1 \cdot 75 \times 10^{-5}$, $K_b = 1 \cdot 77 \times 10^{-5}$), and under these circumstances equation (3.26) predicts that the solution will be very nearly neutral.

If 1 cm^3 of 0·1 mol dm^{-3} hydrochloric acid is added to 1 dm^3 of sodium chloride solution, the pH changes from 7 to about 4 (i.e. a thousandfold change in hydrogen ion concentration). If 1 cm^3 of 0·1 mol dm^{-3} hydrochloric acid is added to 1 dm^3 of ammonium acetate solution, the pH hardly changes. This resistance to pH change on the addition of acid or alkali is known as *buffer action*, and solutions exhibiting this behaviour are called *buffer solutions*.

Buffer solutions almost invariably contain a weak acid and one of its salts with a strong base or a weak base and one of its salts with a strong acid. The solutions are used mainly for resisting changes of pH or for providing solutions of known pH.

If the buffer solution is a mixture of a weak acid and one of its salts, the buffer action occurs in the following way. When hydrogen ions are added, they are removed by combination with the anion of the salt to form the un-ionised acid

$$H_3O^+ + A^- \rightarrow HA + H_2O$$

If, on the other hand, hydroxide ions are added, they are removed by reaction with the free acid to form more salt

$$OH^- + HA \rightarrow A^- + H_2O$$

In the case of a buffer solution which contains a weak base and one of its salts, hydrogen ions are removed by direct combination with the free base to form salt

$$H_3O^+ + B \rightarrow BH^+ + H_2O$$

Hydroxide ions are removed by combination with the cation of the salt to form the un-ionised base

$$OH^- + BH^+ \rightarrow B + H_2O$$

Consider a buffer solution consisting of a weak acid and one of its salts. The dissociation constant of the acid is given by

$$K_a = \frac{[H^+] [A^-]}{[HA]}$$

The term [HA] refers to the concentration of undissociated acid, but if the acid is weak, very little will have ionised and we may put [HA] = [acid], where [acid] is the total concentration of acid in the buffer solution. Similarly, the concentration of A^- ions arising from the dissociation of the acid will be negligible compared with the concentration of A^- ions arising from the salt which will be completely dissociated. Hence $[A^-] = [salt]$. Substituting these approximations in the expression for the dissociation constant, we have

$$K_a = [H^+] \frac{[salt]}{[acid]}$$

or

$$[H^+] = K_a \frac{[acid]}{[salt]}$$

Taking logs

$$\log [H^+] = \log K_a + \log \frac{[acid]}{[salt]}$$

or

$$pH = pK_a + \log \frac{[salt]}{[acid]} \tag{3.27}$$

Equation (3.27) is frequently called the Henderson equation. Its corresponding form for a buffer solution consisting of a weak base and one of its salts is obtained from the expression for the dissociation constant of the base

$$K_b = \frac{[BH^+][OH^-]}{[B]}$$

Making the approximations $[BH^+] = [salt]$ and $[B] = [base]$

$$K_b = [OH^-] \frac{[salt]}{[base]}$$

or

$$[OH^-] = K_b \frac{[base]}{[salt]}$$

whence

$$pOH = pK_b + \log \frac{[salt]}{[base]} \tag{3.28}$$

Remembering that $pH + pOH = pK_w$, eqn. (3.28) may be written in the alternative form

$$pH = pK_w - pK_b - \log \frac{[salt]}{[base]} \qquad (3.29)$$

Owing to the approximations made in the derivation of the Henderson equations (3.27), (3.28) and (3.29), these give good results only in the pH range 4–10. As most of the applications of buffer solutions fall within this region, these equations are quite adequate for most purposes. For buffer solutions of pH < 4 or > 10 rather more complicated equations must be used.

Table 11

BUFFER SOLUTIONS AND RANGES AT $25°C$

Solution	Range
Acetic acid; sodium acetate	3·7–5·6
Potassium dihydrogen phosphate; disodium hydrogen phosphate	5·3–8·0
Boric acid; borax	6·7–9·2

The buffer capacity of a solution may be regarded as the amount of acid or alkali that it can absorb without suffering a change in pH. It is found that buffer solutions have their greatest capacity when the ratio [salt]/[acid] or [salt]/[base] is equal to one. As these ratios diverge from unity the capacity decreases. It is further found that satisfactory buffering can only be obtained when these ratios lie between 10 and $\frac{1}{10}$. Thus the useful ranges of buffer solutions lie between

$$pH = pK_a \pm 1$$

or

$$pOH = pK_b \pm 1$$

Some examples of buffer solutions and their ranges are given in *Table 11*.

THEORY OF ACID–BASE INDICATORS

Indicators for acid–base titrations are invariably weak organic acids or weak organic bases the un-ionised form of which generally exist as two tautomers in equilibrium. Of the two tautomeric forms only one ionises to any appreciable extent. The two tautomers have different colours, and that of the ion produced by the tautomer which ionises is the same as that of the parent tautomer.

Considering an indicator which is itself an acid and denoting the indicator as HIn, where In represents the organic part of the molecule, the equilibria may be written

$$\underbrace{HIn_1}_{Colour\ A} \rightleftharpoons \underbrace{HIn_2 \rightleftharpoons H^+ + In_2^-}_{Colour\ B}$$

where the subscripts 1 and 2 represent the two tautomeric forms of the organic part of the molecule. In acid solution, a large concentration of hydrogen ions exists and this will drive the equilibrium to the left. Colour A is thus the colour of the indicator in acid solution. In alkaline solution, the hydrogen ions produced by the indicator will be removed by combination with hydroxide ions and the above equilibria will lie far over to the right. Colour B is thus the colour of the indicator in alkaline solution. Phenolphthalein may be quoted as an example of this type of indicator, where the equilibria are

Colourless ⇌ Red ⇌ H⁺ + Red

If a substance is to be a good indicator, the concentration of HIn_2 should be very small with respect to that of HIn_1. In this way the alkaline colour will be due almost entirely to In_2^- ions, which are produced to any extent only in alkaline solutions. Similarly, the acid colour due to HIn_1 will not be masked by the alkaline colour due to HIn_2 which is also produced in acid solutions.

The equilibrium constant for the tautomeric equilibrium will be

$$K = \frac{[HIn_2]}{[HIn_1]}$$

It is permissible to use concentrations here, as the concentration of indicator in solution is usually very low and relative activities may be taken as equal to concentrations. As has been mentioned, the constant K must be very small for the substance to be a good indicator and to give a sharp end-point to a titration.

The equilibrium constant for the ionisation will be

$$K' = \frac{[H^+] [In_2^-]}{[HIn_2]}$$

and

$$KK' = \frac{[HIn_2]}{[HIn_1]} \cdot \frac{[H^+] [In_2^-]}{[HIn_2]} = \frac{[H^+] [In_2^-]}{[HIn_1]} = K_i \qquad (3.30)$$

where K_i is known as the *indicator constant*.

In solution, the concentration of HIn_2 will be negligible and the total amount of indicator in the colour B form may be taken as equal to the concentration of In_2^- ions. The total amount of indicator in the colour A form will be equal to the concentration of HIn_1. Hence, if x is the fraction of the indicator in the alkaline-coloured form (colour B in this case), we have

$$x \propto [In_2^-] \quad \text{and} \quad (1 - x) \propto [HIn_1]$$

Substituting in eqn. (3.30), we have

$$K_i = \frac{x}{1 - x} [H^+] \qquad (3.31)$$

Equation (3.31) is known as the *indicator equation*.

Considering an indicator which is itself a weak base and denoting the indicator as InOH, the equilibria may be written

$$\underbrace{In_1OH}_{Colour\ A} \rightleftharpoons \underbrace{In_2OH \rightleftharpoons In_2^+ + OH^-}_{Colour\ B}$$

where the subscripts 1 and 2 once again represent the two tautomeric forms of the organic part of the molecule. In acid solution, the large numbers of hydrogen ions will combine with the hydroxide ions produced by the ionisation of the indicator, and the indicator equilibria will thus lie far to the right. Colour B is therefore that of the indicator in acid solution. Conversely, colour A is the alkaline colour of the indicator.

The equilibrium constant for the tautomeric equilibrium will be

$$K = \frac{[In_2OH]}{[In_1OH]}$$

and once again, if the substance is to be a good indicator, this must be small. The equilibrium constant for the ionisation will be given by

$$K' = \frac{[In_2^+][OH^-]}{[In_2OH]}$$

and

$$KK' = \frac{[In_2OH]}{[In_1OH]} \frac{[In_2^+][OH^-]}{[In_2OH]} = \frac{[In_2^+][OH^-]}{[In_1OH]} \quad (3.32)$$

Now

$$K_w = [H^+][OH^-] \quad \text{and hence} \quad [OH^-] = K_w/[H^+]$$

Substituting in eqn. (3.32)

$$KK' = \frac{[In_2^+]K_w}{[In_1OH][H^+]}$$

and

$$\frac{K_w}{KK'} = \frac{K_w[In_1OH][H^+]}{[In_2^+]K_w} = \frac{[In_1OH][H^+]}{[In_2^+]} = K_i \quad (3.33)$$

where K_i is the indicator constant.

In solution, the concentration of In_2OH will be negligible and the total amount of indicator in the colour B form may be taken as equal to the concentration of In_2^+ ions, while that in the colour A form will be equal to the concentration of In_1OH. If x is the fraction of indicator in the alkaline-coloured form (colour A in this case), we may write

$$x \propto [In_1OH] \quad \text{and} \quad (1 - x) \propto [In_2^+]$$

Substituting in eqn. (3.33), we have

$$K_i = \frac{x}{1 - x} [H^+] \tag{3.34}$$

which is exactly the same result as eqn. (3.31) for indicators which are acids. It must be remembered that x in both cases is the fraction of indicator in the alkaline-coloured form.

In the derivation of the indicator equation we have seen that one condition which must be fulfilled if a substance is to be a good indicator is that the tautomeric equilibrium constant must be small. In addition to this, the equilibrium must be rapidly established and the colour should only be affected by hydrogen ions and not by neutral salts.

THE WORKING RANGE OF INDICATORS

Consider an indicator in a solution which is sufficiently acidic for practically all of the indicator to be in the acid-colour form. If the alkalinity of the solution is gradually increased, the fraction of indicator in the alkaline-coloured form will likewise increase. It has been found that, on average, the human eye can detect no change from the full acid colour until more than 9 per cent of the indicator is in the alkaline-coloured form. There is therefore a critical concentration of hydrogen ions when the eye first detects a change in colour. This concentration may be calculated from the indicator equation (3.34)

$$[H^+] = \frac{1 - x}{x} K_i$$

If 9 per cent of the indicator is in the alkaline-coloured form, $x = 0.09$, and we have

$$[H^+] = \frac{0.91}{0.09} K_i \simeq 10 \, K_i$$

or

$$\log [H^+] = \log 10 + \log K_i$$

and

$$pH = pK_i - 1 \tag{3.35}$$

This equation gives the lowest pH at which the eye can detect a change of colour in the particular indicator.

Similarly, if we consider the indicator in an alkaline solution and gradually decrease the pH, the eye first detects a change from the full alkaline colour when about 9 per cent of the indicator is in the acid-colour form, i.e. when $x = 0.91$. Substituting in the indicator equation

$$[H^+] = \frac{0.09}{0.91} K_i \simeq \frac{1}{10} K_i$$

or

$$\log [H^+] = \log 0.1 + \log K_i$$

and

$$pH = pK_i + 1 \tag{3.36}$$

This equation gives the highest pH at which the eye can detect a change of colour in the indicator. The range of pH over which the indicator appears to change colour is thus approximately two pH units, from $pH = pK_i - 1$ to $pH = pK_i + 1$. This is called the *working range* of the indicator.

The exact extent of the working range will vary from one indicator to another, as the eye has different sensitivity to different colours. The positions of the working ranges of indicators on the pH scale will also vary, depending on the value of the indicator constant as shown by eqn. (3.35) and (3.36). *Table 12* gives the working ranges of some common indicators.

THE USE OF INDICATORS IN ACID–BASE TITRATIONS

The colour of an indicator changes, as mentioned before, over a range of about 2 pH units. If the pH of a solution containing an indicator changes only gradually over its working range, then the resulting colour change will also be gradual. If, however, the pH changes rapidly, the colour change will be sharp and

Table 12

INDICATOR WORKING RANGES

Indicator	Range	pK_i
Bromophenol blue	2·8– 4·6	4·0
Bromocresol green	3·8– 5·4	4·7
Methyl red	4·2– 6·3	5·1
Bromothymol blue	6·0– 7·6	7·0
Phenol red	6·8– 8·4	7·9
Phenolphthalein	8·3–10·0	9·6

(After A. I. Vogel *Quantitative Inorganic Analysis*, 2nd ed., London, Longmans Green, 1951)

provide a good end-point in an acid–base titration. It is clear, then, that the rate of change of pH during a titration, particularly in the neighbourhood of the equivalence point, is of the utmost importance for the accuracy of the titration.

Neutralisation curves for strong, weak and very weak mono basic acids and mono acidic bases have been plotted in *Figure 12(a)* and (*b*); they refer to the neutralisation of 25 cm^3 of 0·1 mol dm^{-3} acids with 0·1 mol dm^{-3} alkali solutions.

The curve for the titration of a strong acid with a strong base is very simply calculated. As the salt formed does not hydrolyse, the pH at the equivalence point (25 cm^3 of alkali added) will be 7 and the pH at other points is calculated from the amount of excess acid or base present. For example, when 22 cm^3 of alkali has been added, there is a 3 cm^3 excess of 0·1 mol dm^{-3} acid in a total volume of 47 cm³. The concentration of the acid will thus be $(0.1 \times 3/47) = 0.0064$ mol dm⁻³. As the acid is strong and thus completely dissociated, the concentration of hydrogen ions in the solution will also be 0.0064 mol dm⁻³, i.e. 6.4×10^{-3} mol dm⁻³, and the pH is thus 2·19.

With a weak or very weak acid and a strong base, the pH of the equivalence point will be given by eqn. (3.20) which gives the

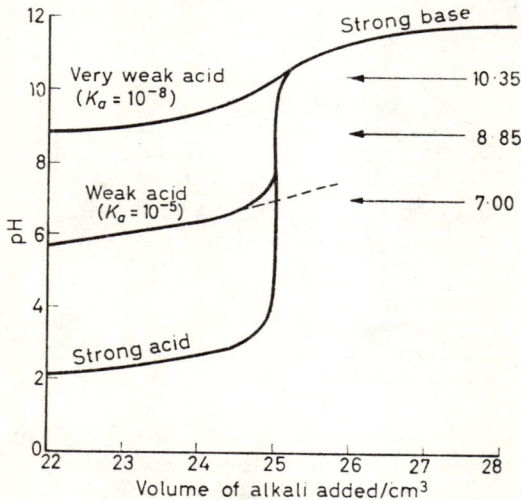

Figure 12a. Neutralisation curves

pH of a solution of a salt of a weak acid and a strong base. The curve on the alkaline side of equivalence will be the same as before, as the excess base is strong. On the acid side of equivalence, the solution is a mixture of a weak acid and one of its salts, thus constituting a buffer solution, the pH of which may be calculated from the Henderson equation (3.27).

Similarly, for the titration of a strong acid with a weak or very weak base, the pH at the equivalence point may be calculated from eqn. (3.22), and the curve on the alkaline side of equivalence from the Henderson equation (3.29).

The curves in *Figure 12* have been calculated for 0·1 mol dm⁻³ solutions, which is a common concentration in acid-base titrations, but it must be remembered that the concentrations of the reactants will influence the shapes of the curves slightly.

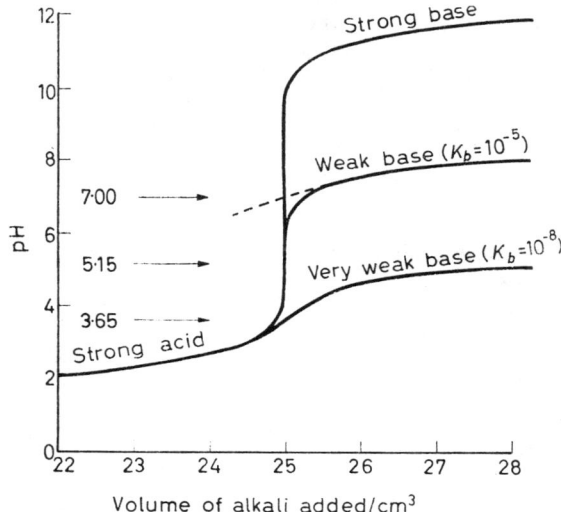

Figure 12b. Neutralisation curves

Consider the titration curve of a strong acid with a strong base. In the region of the equivalence point, the pH changes very rapidly with the volume of alkali added. When 24·95 cm³ of alkali has been added, the pH is 4·00, but this changes to 10·00 on addition of 25·05 cm³ of alkali. This is a change of 6 pH units for two drops of titrant. For such a titration we could choose any indicator with a working range between pH 4 and pH 10. Two drops of titrant at the end-point would then take the pH of the solution right through the working range of the indicator, resulting in its conversion from the full acid colour to the full alkaline. In fact, one drop of titrant changes the pH of the solution by 3 pH units, so that it should be possible to titrate a strong acid with a strong base with a precision of one drop.

In the titration of a weak acid ($K_a = 10^{-5}$) with a strong base, the vertical step in the neutralisation curve is much shorter, covering a range of just over 2 pH units from 7·8 to 10·0, the pH at the equivalence point being 8·85 owing to the hydrolysis of the salt formed. In order to have a sharp end-point for this titration, an indicator with a working range between 7·8 and 10·0 must be selected. Phenolphthalein would be a suitable choice, provided there is not too much carbon dioxide dissolved in the solution, as phenolphthalein is then not satisfactory.

In the titration of a very weak acid ($K_a = 10^{-8}$) with a strong base, the pH at the equivalence point is 10·35 and that of the solution changes only slowly in this region. It can be seen from *Figure 12(a)* that for a change of 2 pH units in the region of the end-point (i.e. from 9·35 to 11·35), the volume of alkali added changes from 23·8 to 26·1 cm^3. Therefore, even if we could choose an indicator with a working range covering the pH of the solution at the end-point, its colour would change gradually over the addition of 2·3 cm^3 of titrant. It is clear, then, that it is not possible to titrate satisfactorily a very weak acid even with a strong base.

When titrating a strong acid with a weak base ($K_b = 10^{-5}$), the pH at the equivalence point is 5·15, and that of the solution changes rapidly in this region from about 4·1 to 6·3. A sharp end-point should be obtainable using an indicator such as methyl red, with a working range between these limits.

With a strong acid and a very weak base ($K_b = 10^{-8}$), there is no sharp change in pH, and a satisfactory titration cannot be carried out.

If a weak acid of $K_a = 10^{-5}$ were titrated with a weak base of $K_b = 10^{-5}$, eqn. (3.26) shows that the pH of the solution at the equivalence point would be 7. The neutralisation curve for this process is obtained by producing the weak acid curve of *Figure 12(a)* to meet the weak base curve of *Figure 12(b)* at pH 7 at the equivalence point. This is indicated by the dotted lines in the diagram. Once again, no sharp change in pH occurs and a satisfactory titration is not possible.

SOLUBILITY PRODUCTS AND SOLUBILITIES

When a saturated solution of an electrolyte is in contact with the solid solute, there exists an equilibrium between the ions in the solution and the solid. In the case of a uni-univalent electrolyte, this equilibrium may be represented by

$$MA(s) \rightleftharpoons M^+ + A^-$$

and the equilibrium constant, K_s, is given by

$$K_s = \frac{a_{M^+} \times a_{A^-}}{a_{MA}}$$

It has been pointed out above that the relative activity of a pure liquid is taken as unity. In the same way, the relative activity of a pure solid is unity, and thus the above expression becomes

$$K_s = a_{M^+} \times a_{A^-} \tag{3.37}$$

The constant K_s is known as the *activity solubility product*.

The significance of eqn. (3.37) is that the product of the relative activities of the ions M^+ and A^- in a saturated solution of MA must be constant for a given temperature. This statement holds true irrespective of whether the ions arise solely from the substance MA or whether any contribution is made to their relative activities by the presence of other electrolytes in the solution.

Equation (3.37) may be written in terms of concentrations and activity coefficients

$$K_s = (c_{M^+}c_{A^-}) \times (y_{M^+}y_{A^-}) \tag{3.38}$$

The solubility product principle expressed by eqn. (3.37) and (3.38) applies to any saturated solution but it is of practical interest only in the cases of sparingly soluble salts.

Consider a saturated solution of a sparingly soluble salt, MA, in water, the solution being in contact with the pure solid solute. As the salt is only sparingly soluble, the solution will be extremely dilute and the ionic activity coefficients may be taken as unity. Under these circumstances, eqn. (3.38) reduces to

$$K_s = c_{M^+} \times c_{A^-} \qquad (3.39)$$

If the salt is completely dissociated in solution, $c_{M^+} = c_{A^-} = c_{MA}$, where c_{MA} is the concentration of the salt in the saturated solution. Equation (3.39) may thus be written

$$K_s = c_{MA}^2$$

or

$$c_{MA} = \sqrt{K_s}$$

The solubility of the salt is thus given as the square root of the solubility product. Great care should be exercised in applying this result. It depends on the assumption that the salt is completely dissociated in solution. If M^+ ions associate with A^- ions to any extent, they form what are virtually un-ionised MA molecules in solution, and the true solubility of the salt will be greater than that given by the solubility product.

We must now examine the effects of added electrolytes on a saturated solution of the sparingly soluble salt MA in contact with solid MA. Suppose in the first instance that an electrolyte which contains neither M^+ ions nor A^- ions is added to the solution. Such an electrolyte is sometimes referred to as an indifferent electrolyte. The effect of adding it will be to increase the ionic strength of the solution. This will result in a decrease in the ionic activity coefficients, y_{M^+} and y_{A^-}, and in order that eqn. (3.38) should remain true, the concentrations of the ions M^+ and A^- must increase. The addition of an indifferent electrolyte will, in the first instance, increase the solubility of the salt MA. It will be remembered, however, that as ionic strength increases ionic activity coefficients decrease to a minimum and then increase again. There will come a point, then, at very high concentrations of indifferent electrolyte when the solubility of MA may be decreased. In most practical applications, however, this stage is never reached.

When the added electrolyte has an ion in common with the salt MA, the position is somewhat different. Suppose that the added electrolyte contains the ion A^-. As the ionic strength of the solution increases, the activity coefficient term, $(y_{M^+}y_{A^-})$, in eqn. (3.38) will decrease. This decrease, however, will be outweighed by the increase in the c_A- term due to the added A^- ions. Thus, in order for the equation to hold, the c_{M+} term must decrease, and this is reflected as a decrease in the solubility of the salt MA. This is sometimes called 'the common ion effect'.

It must be remembered that, at very high concentrations of added electrolyte, the term $(y_{M+} \, y_A-)$ will increase and thus support the increase in the c_A- term. This will result in a greater decrease of c_{M+} and, correspondingly, in the solubility of the salt MA. Once again, this stage is not reached in most cases of practical interest.

In addition to the factors discussed above which affect eqn. (3.38), there is a further effect which may influence the solubility of the salt MA. If any of the added ions can form a complex with either the M^+ or A^- ions, the result will be an increase in the solubility of the salt. If, for example, the salt under consideration were silver chloride, the equilibrium in the absence of added salts would be

$$AgCl(s) \rightleftharpoons Ag^+ + Cl^-$$

Suppose now that a cyanide were added to the solution. Cyanide ions would react with silver ions to form argentocyanide ions

$$Ag^+ + 2CN^- \rightarrow Ag(CN)_2^-$$

This reaction would disturb the first equilibrium and more silver chloride would dissolve. Complexing tends to increase solubility irrespective of whether the added electrolyte has a common ion or not. The effects of complexing may be observed at quite moderate concentrations well below those required for a change in effect of the activity coefficients.

The effects of added salts within the region of practical interest may be summarised as follows.

When the added electrolyte is indifferent, the solubility of the salt increases with increasing concentration of added electrolyte. This effect may be enhanced by complex formation.

When the added electrolyte has a common ion, the solubility of the salt will at first decrease with increasing concentration of added electrolyte. This decrease may pass through a minimum and become an increase at moderate concentrations due to **complex formation.**

4

REVERSIBLE ELECTRODE POTENTIALS

HALF CELLS

IF A METAL is partially immersed in a solution of its ions, a separation of charge is found to occur and a potential difference is established between the metal and the solution. Some of the metal atoms lose electrons and pass into solution as metal ions:

$$M \rightarrow M^+ + e$$

This process would result in the accumulation of the liberated electrons in the metal which would become negatively charged with respect to the solution.

Similarly, some of the metal ions in solution will abstract electrons from the metal and deposit as metal atoms:

$$M^+ + e \rightarrow M$$

This process would lead to a deficit of electrons in the metal which would then become positively charged with respect to the solution.

If the first reaction occurred more rapidly than the second, the metal would acquire a net negative charge which would make it more difficult for positive ions to leave the metal and thus retard the rate of the reaction. Again, the negative charge on the metal would attract the positive ions in solution, thus accelerating the second reaction. In this way the rates of the two reactions become equal and equilibrium is established.

If the second reaction occurred more rapidly initially, the metal would acquire a net positive charge, thus accelerating the first reaction and retarding the second, leading once again to the establishment of equilibrium.

The equilibrium between these two processes may be represented by

$$M \rightleftharpoons M^+ + e$$

the final potential adopted by the metal depending upon the position of this equilibrium which, in turn, depends upon the relative activities of the species involved. The potential is known as the *electrode potential* of the metal, and if the system is in thermodynamic equilibrium, as the *reversible electrode potential* of the metal.

The concept of electrode potentials is not confined to metals, and electrodes can be devised in which gases are in equilibrium with ions in solution. For example, it is possible to bubble chlorine around an electrode of platinised platinum immersed in a solution containing chloride ions. The chlorine gas is adsorbed onto the surface of the platinum and enters into equilibrium with the chloride ions in the solution

$$\tfrac{1}{2}Cl_2 + e \rightleftharpoons Cl^-$$

In this case, the electrons are abstracted from, or accumulate on, the inert platinum electrode which takes no chemical part in the reaction but acts merely as an electrical conductor.

It will be noticed that in the examples quoted above the equilibria involve two opposing reactions, an oxidation reaction and a reduction reaction. This must be so, as the reactions involve a loss or a gain of electrons. The equilibria exist, therefore, between the oxidised and reduced forms of the electrode systems. In a metal electrode, the metal itself is the reduced form (possessing more electrons) and the metal ions are the oxidised form (possessing less electrons). In the case of the chlorine electrode, the chlorine gas is the oxidised form and the chloride ions are the reduced form.

Such systems are known as *half cells*, and the potentials which they develop are commonly called the electrode potentials of the systems involved. There is, however, another type of half cell in which both the oxidised and reduced forms of the system exist in solution. An example of this type is provided by a platinum electrode immersed in a solution containing ferrous and ferric ions. The equilibrium involved here is

$$Fe^{3+} + e \rightleftharpoons Fe^{2+}$$

the platinum electrode supplying or accepting electrons as it does in the case of a chlorine electrode. Although there is no fundamental difference between this last half cell and the first two examples given, an electrode in which the oxidised and reduced forms both exist in solution has come to be called a *redox electrode*, and its potential is known as a *redox potential.*

We have seen already that the magnitude of the potential developed by a half cell, be it an electrode potential or a redox potential, will depend upon the activities of the species involved in the equilibrium. Before going on to consider the relationship between potential and activity, it will probably be advantageous to give further consideration to the concept of activity.

ACTIVITY OF GASES

In Chapter 3 it was shown that the relative activities of pure liquids and pure solids are taken as unity. These relative activities which we have considered are really the activities of substances relative to their activity in some arbitrarily chosen standard state. The standard state which is usually chosen for solids and liquids is the form which is stable at 25°C and 1 atm. pressure. Strictly then, the relative activities of pure solids and liquids are unity only under these conditions. On most occasions in electrochemistry, however, it will be sufficient to consider the relative activities of pure solids and liquids as equal to unity, and unless specifically stated otherwise it will be assumed that this is the case.

[As we shall have to mention relative activity quite frequently from now on the usual practive of abbreviating the term simply to *activity* will be adopted in the future].

With gases, the position is slightly different. It will be remembered that activity may be regarded as a corrected form of concentration which allows for molecular or ionic interaction.

For a gas it is often convenient to measure its concentration by its partial pressure. The activity of a gas, then, may be regarded as a corrected partial pressure, where the correction allows for the non-ideal behaviour of the gas. If the gas behaves ideally, its activity will be equal to its partial pressure (measured in atmospheres). At low pressures of less than several atmospheres, say, most gases behave very nearly ideally and under these conditions it is adequate to put the activity of the gas equal to its partial pressure. Many electrochemical observations are made at a pressure of 1 atm and if the partial pressure of any gas involved in a gas electrode is equal to the pressure of the atmosphere it will be sufficiently accurate to take its activity as unity.

To sum up then, we may take the activity of a pure solid to be unity and the activity of a gas as equal to its partial pressure at pressures of less than a few atmospheres.

When a substance has an activity of unity, it is said to be in its *standard state* of unit activity. This term arises from thermodynamics and its origin need not trouble us here, but we must consider how it applies to electrodes.

If all the components involved in the equilibrium in a half cell are at unit activity, they are all in their standard states and the half cell is said to be a standard electrode. For example, if a piece of pure zinc is immersed in a solution of zinc ions at unit activity, it constitutes a standard zinc electrode. If, in a chlorine half cell, the chlorine gas is at a partial pressure of 1 atm and the chloride ions in solution are at unit activity, the system constitutes a standard chlorine electrode.

The potentials of electrodes which are in their standard states are known as the *standard electrode potentials* of the systems. The potential of a standard zinc electrode, for example, is called the standard electrode potential of zinc.

VARIATION OF ELECTRODE POTENTIAL WITH ACTIVITY

It has been pointed out already that at an electrode an equilibrium exists between the oxidised and reduced forms of a system.

Equation (4.1) gives the relationship between electrode potential and the activities of the oxidised and reduced forms of the electrode system

$$E = E^\circ + \frac{RT}{zF} \ln \frac{(\text{activity of oxidised form})}{(\text{activity of reduced form})} \qquad (4.1)$$

In this equation, E = electrode potential, E° = standard electrode potential, R = gas constant, T = temperature, z is the number of electrons involved in the electrode reaction. F is Faraday's constant.

Equation (4.1) will be proved below, but for the time being it will suffice to see how it applies to various half cells. The potential of the ferrous–ferric redox system will be given by

$$E = E^\circ + \frac{RT}{F} \ln \frac{a_{Fe^{3+}}}{a_{Fe^{2+}}} \qquad (4.2)$$

In this case, E will be called the redox potential and E° will be the standard redox potential, as the oxidised and reduced forms both exist in solution. Also, $z = 1$, as can readily be seen from the fact that only one electron is involved in the electrode reaction

$$Fe^{3+} + e \rightleftharpoons Fe^{2+}$$

If we consider a more complex electrode reaction such as that between chromium (III) and dichromate ions in acid solution, the equilibrium at the electrode will be

$$Cr_2O_7^{2-} + 14H^+ + 6e \rightleftharpoons 2Cr^{3+} + 7H_2O$$

The number of electrons involved in the equilibrium is six, so that $z = 6$, and making the approximation that the activity of water will be unity in dilute solution, eqn. (4.1) takes the form

$$E = E^\circ + \frac{RT}{6F} \ln \frac{a_{Cr_2O_7^{2-}} \times a_{H^+}^{14}}{a_{Cr^{3+}}^2} \qquad (4.3)$$

It will be noted that the activities of all the species involved in the equilibrium are included in the logarithmic term.

In two particular cases, eqn. (4.1) may be reduced to a slightly simpler form. The first is that of a metal electrode in equilibrium with its positively charged ions. The equilibrium could be

$$M \rightleftharpoons M^{z+} + ze$$

and applying eqn. (4.1), we have

$$E = E^\circ + \frac{RT}{zF} \ln \frac{a_+}{a_M} \tag{4.4}$$

where a_+ is the activity of the positively charged ions in the solution and a_M that of the metal electrode. As the latter is a pure solid, its activity is unity, and eqn. (4.4) becomes

$$E = E^\circ + \frac{RT}{zF} \ln a_+ \tag{4.5}$$

The second case is where an electrode is in equilibrium with negatively charged ions in solution. This could well be the case with a gas electrode such as a chlorine electrode. Suppose for the general case the equilibrium to be represented as

$$X + ze \rightleftharpoons X^{z-}$$

Applying eqn. (4.1)

$$E = E^\circ + \frac{RT}{zF} \ln \frac{a_X}{a_-} \tag{4.6}$$

where a_- is the activity of the negatively charged ions in solution and a_X that of the gas. At atmospheric pressure, the activity of the gas will be unity and eqn. (4.6) becomes

$$E = E^\circ + \frac{RT}{zF} \ln \frac{1}{a_-}$$

or

$$E = E^\circ - \frac{RT}{zF} \ln a_- \tag{4.7}$$

Equations (4.5) and (4.7) were originally deduced in a rather different way by Nernst and were at one time commonly called the Nernst equations.

ELECTROCHEMICAL CELLS AND CELL REACTIONS

When two half cells are combined, an *electrochemical cell* or a *galvanic cell* is obtained. For example, a zinc electrode could be combined with a chlorine electrode, as illustrated in *Figure 13*. The cell may be represented diagrammatically as

$$Zn/ZnCl_2/Cl_2; Pt$$

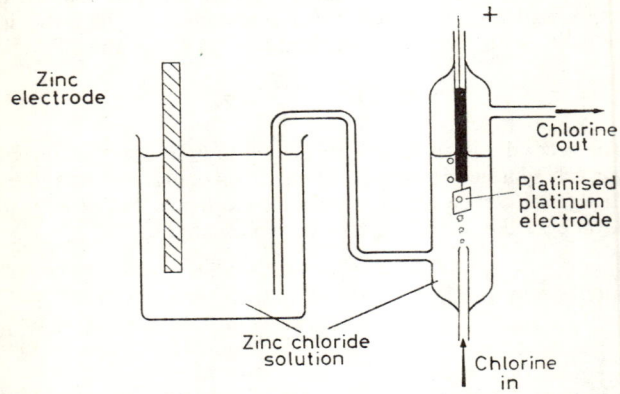

Figure 13. Zinc–chlorine cell

The electrolyte is zinc chloride solution and is the same throughout the cell, the chlorine electrode being in equilibrium with the chloride ions and the zinc electrode with the zinc ions. In such a cell it would be found experimentally that the potential of the zinc electrode was more negative than that of the chlorine electrode. The zinc is thus called the negative pole of the cell, whilst the chlorine is the positive pole. If the electrodes were connected with an external circuit, electrons would flow round it from the zinc electrode to the chlorine electrode. As soon as electrons leave the zinc, the electrode equilibrium is disturbed and more electrons are required to try to maintain the zinc electrode at its equilibrium potential. These electrons are provided by the ionisation of more zinc atoms from the electrode to form zinc ions in solution. There is thus a net chemical reaction occurring at the zinc electrode which is

$$Zn \rightarrow Zn^{2+} + 2e$$

As electrons arrive at the chloride electrode from the external circuit, the potential of the electrode becomes more negative than the equilibrium potential and a reaction occurs which consumes electrons in order to try and restore the electrode to its equilibrium potential:

$$Cl_2 + 2e \rightarrow 2Cl^-$$

The total reaction occurring in the cell is the sum of the two electrode reactions and may be written

$$Zn + Cl_2 \rightarrow Zn^{2+} + 2Cl^-$$

This is known as the *cell reaction*, and essentially the cell is converting the chemical energy released by this reaction into electrical energy.

Once the polarity of an electrochemical cell is known, the cell reaction may always be deduced by applying the following principles. The cell reaction will occur when the electrode equilibrium is disturbed by connecting the cell to an external circuit, so that the cell drives current round it. Electrons will flow from the more negative electrode round the circuit to the positive pole of the cell. To sustain this current, a reaction must occur at the negative pole which is capable of supplying electrons to the external circuit and one must occur at the positive pole which is capable of consuming the electrons which arrive from the external circuit. This means that the reaction occurring at the negative pole is always an oxidation reaction and that the one occurring at the positive pole is always a reduction reaction.

Another example of an electrochemical cell is the Daniell cell which consists of a zinc half cell and a copper half cell. This could be arranged as shown in *Figure 14*. The cell comprises a

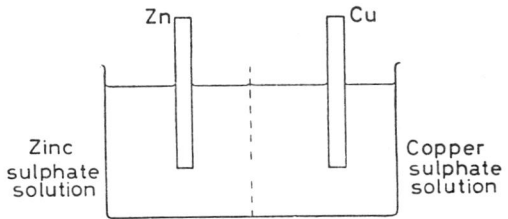

Figure 14. Daniell cell

zinc electrode partially immersed in a solution of zinc sulphate and a copper electrode partially immersed in a solution of copper sulphate. The two solutions are separated by a permeable membrane or sintered-glass partition. In this cell the zinc electrode is usually more negative than the copper electrode. The zinc is thus the negative pole of the cell and the copper the positive. This does not necessarily mean that the copper electrode is positive with respect to the solution but merely that its potential is more positive than that of the zinc electrode. This cell could be represented diagramatically as

$$Zn/ZnSO_4 \mid CuSO_4/Cu$$

If the cell were allowed to supply current to an external circuit, a reaction furnishing electrons would occur at the zinc electrode (negative pole) and a reaction consuming electrons would occur at the copper electrode (positive pole):

$$Zn \rightarrow Zn^{2+} + 2e$$

and

$$Cu^{2+} + 2e \rightarrow Cu$$

The overall cell reaction is thus

$$Zn + Cu^{2+} \rightarrow Zn^{2+} + Cu$$

Once again, oxidation occurs at the negative pole (Zn) and reduction at the positive pole (Cu). This is the familiar reaction of zinc displacing copper from solution, and it is used in the Daniell cell to supply electrical energy.

The electromotive force of an electrochemical cell is equal to the difference between the potentials of the two electrodes comprising the cell.

If it so happens that both of the electrode systems comprising the cell are in their standard states, then the cell as a whole is in its standard state and the e.m.f. of the cell is called the standard e.m.f. It is, of course, equal to the difference between the standard electrode potentials of the individual electrodes.

LIQUID JUNCTION POTENTIALS

In the Daniell cell, some diffusion will occur at the interface between the copper sulphate solution and the zinc sulphate solution. Ignoring the sulphate ions for the moment, zinc ions will diffuse into the copper solution and copper ions into the zinc solution. If the rates of diffusion of both ions are the same, then equal amounts of positive charge will be transferred across the interface in opposite directions, so that there will be no net transfer of charge. If, as is more usually the case, one ion diffuses more rapidly than the other, there will be a separation of charge across the interface and a potential difference will be established. Suppose, with reference to *Figure 14*, that the copper ions diffuse more rapidly from right to left than the zinc ions diffuse from left to right. There will be a net transfer of positive charge across the boundary and the left-hand side of the interface will become positive with respect to the right-hand side. The copper ions are now diffusing from a more negative to a more positive potential, and as copper ions are positively charged, the electric field produced by the difference of potential will tend to retard them. On the other hand, the zinc ions are now diffusing from a more positive towards a more negative potential, which will accelerate the slower-moving zinc ions. Eventually the rates of diffusion of the two ions become equal and a steady state is reached with a steady potential difference across the interface. Usually there will also be diffusion of the sulphate ions across the boundary, which will increase or decrease the equilibrium potential across the interface according to the direction of diffusion.

In general, there is always a potential difference established across the interface between two dissimilar electrolytes, and these potentials are known as *liquid junction potentials* or sometimes *diffusion potentials*. The potential always acts in such a way as to retard the more rapidly diffusing ions and accelerate the more slowly diffusing ions, whether they be cations or anions. In this way, equilibrium is soon reached and a steady liquid junction potential is established. Its magnitude will depend upon the speeds of the ions involved and upon the charge which they carry.

Liquid junction potentials do not usually exceed $0 \cdot 1V$, but they are a nuisance in experimental work, as they are difficult to reproduce and it is desirable that they should be eliminated. Unfortunately, it is not possible to eliminate them entirely but they can be reduced to a negligible value by using a *salt bridge*. This consists of a tube containing either saturated potassium chloride or saturated ammonium nitrate solution, and it is used as shown in *Figure 15*. The ends of the salt bridge tube are usually plugged with cotton wool to prevent excessive diffusion. By using a salt bridge one liquid junction is replaced by two. At each junction, the vast majority of any current is carried by the electrolyte of the salt bridge, as this is usually so much more concentrated than the electrode solutions. It can be shown that, in the special case of a junction between two solutions of a uni-univalent electrolyte of activities a_1 and a_2, the liquid junction potential, E_l, is given by

$$E_l = (t_- - t_+) \frac{RT}{F} \ln \frac{a_2}{a_1} \tag{4.8}$$

where t_+ and t_- are the transport numbers of the cation and anion, respectively.

When using a salt bridge, it may be considered that most of the diffusion is done by the bridge electrolyte, and eqn. (4.8) can be applied. The reason for choosing potassium chloride or ammo-

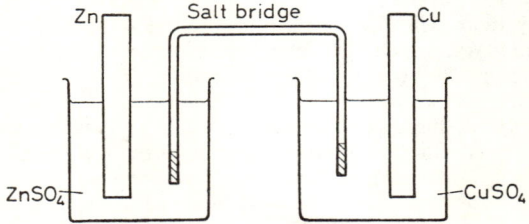

Figure 15. Daniell cell with salt bridge

nium nitrate as suitable bridge electrolytes lies in the fact that in both these salts the transport numbers of the ions are very close to 0·5. It can be seen from eqn. (4.8) that, in this case, the liquid junction potential will be very nearly zero. We have thus replaced one liquid junction potential by two, acting in opposite directions and having values which are very nearly zero.

The cell illustrated in *Figure 15* is usually represented by the diagram

$$Zn/ZnSO_4//CuSO_4/Cu$$

the double stroke in the centre representing the elimination of any liquid junction potential.

MEASUREMENT OF E.M.F.

The e.m.f. of electrochemical cells are measured by comparing the e.m.f. of the cell in question with that of a comparison cell. Such cells are frequently called standard cells. The use of the term standard here does not refer to the standard state but merely implies that the e.m.f. of the standard cell is already accurately established. The most common standard cell in use is the Weston cell (*Figure 16*). It has an e.m.f. of 1·01859 V at 20°C, and this voltage will remain constant over a period of years.

The e.m.f. to be measured is compared with that of the Weston cell by using a potentiometer circuit (*Figure 17*). A bridge wire, *AB*, is connected to an accumulator or some other

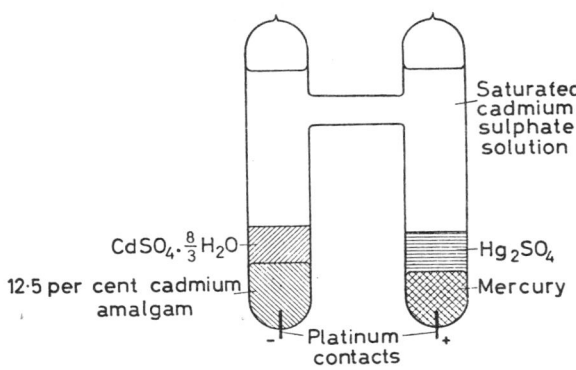

Figure 16. Weston cell

99

d.c. source. The wire takes a small current, and if the connecting leads have a low resistance, the voltage drop along the length of the wire will be equal to the potential difference between the plates of the accumulator. As shown in the diagram, the positive pole of the latter is connected to A and the negative pole to B. If the accumulator potential difference is 2 V, point A will be 2 V positive with respect to point B. The bridge wire should be uniform, so that a point half-way between A and B will be 1 V negative with respect to A and 1 V positive with respect to B.

The positive pole of the Weston cell is also connected to A, and the negative pole is connected by a switch K to a galvanometer which is connected to a sliding contact on the bridge wire. The cell to be measured is connected in parallel with the Weston cell between point A and switch K.

The Weston cell is brought into circuit by placing the switch in the position shown in the diagram and the sliding contact is adjusted until the galvanometer shows no deflection. Suppose the

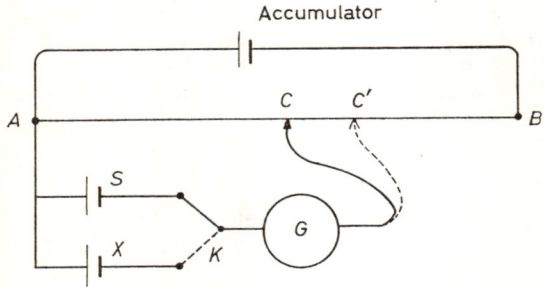

Figure 17. Potentiometer circuit

position of the sliding contact is then at C as shown. If the galvanometer shows no deflection, no current can be passing through the Weston cell, and this means that the e.m.f. of the cell must be exactly balanced by the voltage tapped off the bridge wire between points A and C. If the e.m.f. of the Weston cell is E_s, then we may write

$$E_s \propto AC$$

The unknown cell X is now brought into circuit by moving the switch to the position shown by the dotted line, and a new balance point, C', is found. If the e.m.f. of cell X is E_x, then

$$E_x \propto AC'$$

and

$$\frac{E_x}{E_s} = \frac{AC'}{AC}$$

or

$$E_x = E_s \frac{AC'}{AC} \qquad (4.9)$$

It often happens that the polarity of the cell X is unknown, and it is therefore not possible to know which electrode should be connected to A. If the cell is connected correctly, its e.m.f. will be in opposition to the voltage supplied by the bridge wire and a balance point will be found. If, however, the cell is connected the wrong way round, its e.m.f. will be in series with the bridge voltage and it will be impossible to achieve a condition of zero current through the galvanometer. In this case no balance point will be found.

There are thus *two* pieces of information to be derived from a potentiometric measurement. Firstly the *e.m.f. of the cell* is determined, and secondly the *polarity* of the cell may be deduced.

SIGN CONVENTIONS

As with all scientific information it is desirable that the two pieces of information derived from a potentiometric measurement should be recorded in a standard manner. There are several possible ways of doing this and in the past there were two methods used concurrently; one used largely by American chemists and the other by European chemists. The generally accepted method which is currently used is that proposed by the International Union of Pure and Applied Chemistry (I.U.P.A.C.) in 1953 and has come to be known as the international sign convention.

On the international convention the cell is represented by a cell diagram and the e.m.f. of the cell is written as the potential of the electrode on the right-hand side of the diagram with respect to that of the electrode on the left. The e.m.f. of the cell may thus turn out to be positive or negative depending on the way in which the diagram is written and the relative potentials of the two constituent electrodes.

To illustrate this statement, consider the Daniell cell shown in *Figure 15*. Suppose that as a result of a potentiometric measurement the e.m.f. of the cell was found to be 1 V with the copper electrode as the positive and the zinc electrode as the negative pole of the cell. It could be said that the copper electrode was 1 V positive with respect to the zinc electrode or that the zinc electrode was 1 V negative with respect to the copper electrode. If the cell were represented by the diagram,

$$Zn/ZnSO_4//CuSO_4/Cu$$

the potential of the right-hand electrode (Cu) would be $+1$ V with respect to the left-hand electrode (Zn) and the e.m.f. of the cell would be given by $E_{cell} = +1$ V.

Alternatively, if the cell were represented by the diagram,

$$Cu/CuSO_4//ZnSO_4/Zn$$

the potential of the right-hand electrode (Zn) would be -1 V with respect to the left-hand electrode (Cu) and the e.m.f. of the cell would be given by $E_{cell} = -1$ V.

A typical statement of the result of a potentiometric measurement might thus be

$$Mg/Mg(NO_3)_2//Pb(NO_3)_2/Pb \qquad E = +2·24 \text{ V}$$

This would be interpreted as meaning that the e.m.f. of the cell was 2·24 V and that the lead electrode was the positive pole of the cell. Having established the polarity of the cell in this way the cell reaction may be deduced as discussed in the section on cell reactions. Another method of deducing the cell reaction will be discussed later when the thermodynamics of cells have been considered but the above method is recommended as it is founded on more basic principles.

It is to be noted that on the basis of the above sign convention the e.m.f. of a cell is always obtained by subtracting the potential of the left-hand electrode in the diagram from that of the right hand electrode. Thus

$$E_{cell} = E_{right} - E_{left} \qquad (4.10)$$

REVERSIBLE ELECTRODE POTENTIALS
THE HYDROGEN SCALE OF ELECTRODE POTENTIALS

All ordinary measurements of the e.m.f. of cells give the potential of one electrode with respect to the other. As has been shown above with the Daniell cell it could only be said that the copper electrode was 1 V positive with respect to the zinc electrode or that the zinc electrode was 1 V negative with respect to the copper electrode, but the absolute magnitude of the potential of either electrode could not be established. It is not possible to measure the potential of a single electrode, as the half cell would have to be connected to the measuring device, and this presents the problem of making electrical contact with the solution without introducing a second metal–solution interface. Practical measurements, then, always yield a difference between two individual electrode potentials.

If we wish to assign particular values to the various electrode potentials, we can adopt an arbitrary zero of potential. The standard now adopted is that originally due to Nernst who suggested that *the potential of a standard hydrogen electrode be taken as zero at all temperatures.* In this way, all other electrode potentials can be referred to the standard hydrogen electrode. If, for example, it is found that the e.m.f. of a cell in which a certain electrode is coupled with a standard hydrogen electrode is x V, with the hydrogen electrode as the negative pole, it may now be said that the potential of the electrode in question is $+ x$ V on the hydrogen scale or, more briefly, that it is $+ x$ V. If the hydrogen electrode had been the positive pole of the cell, the electrode in question would have been assigned a potential of $- x$ V.

The potential of any electrode is thus expressed with respect to a standard hydrogen electrode at the same temperature, and the latter is established as a *primary reference electrode.*

Another way óf saying this is that the potentials of half cells are defined in terms of the e.m.f. of a complete cell, formed by combining the half cell with a standard hydrogen electrode, this electrode being written down as the left-hand electrode of the cell. Thus the potential of the half cell Zn^{2+}/Zn means the e.m.f. of the cell

ELEMENTARY ELECTROCHEMISTRY

$$Pt;H_2(a = 1)/H^+(a = 1)//Zn^{2+}/Zn$$

The e.m.f. of this cell is given as the potential of the zinc electrode (right-hand electrode) with respect to that of a standard hydrogen electrode (left-hand electrode).

It is to be noted that the abbreviation for the above cell is obtained by simply deleting the hydrogen electrode to leave Zn^{2+}/Zn.

Table 13

STANDARD ELECTRODE POTENTIALS AT $25°C$

	V		V
Li^+/Li	$-3\cdot045$	Sn^{2+}/Sn	$-0\cdot136$
K^+/K	$-2\cdot925$	Pb^{2+}/Pb	$-0\cdot126$
Ca^{2+}/Ca	$-2\cdot87$	Cu^{2+}/Cu	$+0\cdot337$
Na^+/Na	$-2\cdot714$	Ag^+/Ag	$+0\cdot7991$
Mg^{2+}/Mg	$-2\cdot37$	Au^{3+}/Au	$+1\cdot50$
Al^{3+}/Al	$-1\cdot66$	OH^-/O_2	$+0\cdot401$
Zn^{2+}/Zn	$-0\cdot763$	I^-/I_2	$+0\cdot5355$
Fe^{2+}/Fe	$-0\cdot440$	Br^-/Br_2	$+1\cdot0652$
Cd^{2+}/Cd	$-0\cdot403$	Cl^-/Cl_2	$+1\cdot3595$
Tl^+/Tl	$-0\cdot3363$	F^-/F_2	$+2\cdot65$

(After W. M. LATIMER *Oxidation Potentials*, 2nd ed., New York, Prentice-Hall, 1952)

Although we have taken the potential of the standard hydrogen electrode to be zero at all temperatures, this does not necessarily mean that the potential of this electrode does not vary with temperature. In fact, it probably varies considerably, but this will make no difference to the scale of electrode potentials at any particular temperature. We are merely imposing the arbitrary value of zero on the temperature coefficient of the potential of the standard hydrogen electrode.

Table 13 gives some values of standard electrode potentials on the hydrogen scale.

SUBSIDIARY REFERENCE ELECTRODES

In practice, it is not always easy or convenient to set up a hydrogen electrode. Such electrodes cannot be used in solutions containing reducible substances and they are subject to poisoning by such substances as poison heterogeneous catalysts. Although the standard hydrogen electrode remains the primary reference electrode upon which all potential measurements are based, subsidiary reference electrodes are often employed.

A subsidiary reference electrode can be any electrode the potential of which with respect to a standard hydrogen electrode has previously been accurately determined. Although any electrode could be chosen for this purpose, there is the practical consideration that it should give consistently reproducible results. This limitation restricts the choice, and the most common form of subsidiary reference electrode in use today consists of a metal in contact with a solution which is saturated with a sparingly soluble salt of the metal and which also contains an additional salt with a common anion. Examples are provided by the following electrode systems:

$$Ag/AgCl(s), KCl \text{ aq.}$$

$$Hg/Hg_2SO_4(s), K_2SO_4 \text{ aq.}$$

$$Hg/Hg_2Cl_2(s), KCl \text{ aq.}$$

In each case, the potential of the electrode is governed by the activity of the anion in the solution.

Consider a silver–silver chloride electrode containing potassium chloride, the activity of the chloride ions in the solution being a_{Cl^-}. The potential of the silver electrode depends directly upon the activity of the silver ions, a_{Ag^+}, in the solution as predicted by eqn. (4.5) which takes the form

$$E = E_{Ag}^\circ + \frac{RT}{F} \ln a_{Ag^+} \tag{4.11}$$

where E_{Ag}° is the standard potential of the silver electrode. According to the solubility product principle, the product of the activities of the silver ions and the chloride ions in the solution will be equal to the solubility product of silver chloride, K_{AgCl}.

Thus

$$a_{Ag^+} \times a_{Cl^-} = K_{AgCl}$$

or

$$a_{Ag^+} = \frac{K_{AgCl}}{a_{Cl^-}}$$

Substituting in eqn. (4.11)

$$E = E_{Ag}{}^\circ + \frac{RT}{F} \ln \frac{K_{AgCl}}{a_{Cl^-}}$$

or

$$E = E_{Ag}{}^\circ + \frac{RT}{F} \ln K_{AgCl} - \frac{RT}{F} \ln a_{Cl^-}$$

$$(4.12)$$

The term $(RT/F) \ln K_{AgCl}$ will be constant at constant temperature and may be combined with $E_{Ag}{}^\circ$ to give a new constant, denoted $E_{AgCl}{}^\circ$, which may be regarded as the standard potential of the silver–silver chloride electrode. Equation (4.12) thus becomes

$$E = E_{AgCl}{}^\circ - \frac{RT}{F} \ln a_{Cl^-} \qquad (4.13)$$

The same conclusion could have been obtained by the application of eqn. (4.1) to the electrode equilibrium, which may be written

$$Ag \rightleftharpoons Ag^+ + e$$

and

$$Ag^+ + Cl^- \rightleftharpoons AgCl(s)$$

Or, combining the two stages

$$Ag + Cl^- \rightleftharpoons AgCl(s) + e$$

106

Applying eqn. (4.1)

$$E = E_{AgCl}{}^\circ + \frac{RT}{F} \ln \frac{a_{AgCl}}{a_{Ag} \times a_{Cl^-}} \qquad (4.14)$$

As the silver and silver chloride are pure solids, their activities will be equal to unity and eqn. (4.14) becomes

$$E = E_{AgCl}{}^\circ + \frac{RT}{F} \ln \frac{1}{a_{Cl^-}}$$

or

$$E = E_{AgCl}{}^\circ - \frac{RT}{F} \ln a_{Cl^-}$$

It is interesting to note, however, that

$$E_{AgCl}{}^\circ = E_{Ag}{}^\circ + \frac{RT}{F} \ln K_{AgCl}$$

Precisely similar arguments will apply to the other two subsidiary reference electrodes. The potentials of each are governed by the activities of the sulphate ions and chloride ions, respectively.

It is to be noted in passing that electrode systems of this type, sometimes called ' electrodes of the second kind ', provide a useful method of setting up electrodes which are reversible to anions. In some cases, this would prove difficult or even impossible by means of a simple electrode system.

Probably the most common subsidiary reference electrode is the mercury–mercurous chloride electrode which is known as the *calomel electrode*. When the solution is saturated with respect to potassium chloride, it is called the *saturated calomel electrode*. This is commonly used for ordinary measurements but for accurate work a calomel electrode containing 1 mol dm^{-3} KCl is preferred as its potential is less temperature-dependent. A common form of calomel electrode is illustrated in *Figure 18*.

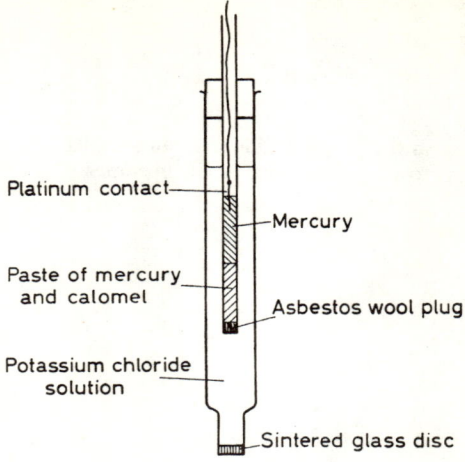

Platinum contact

Mercury

Paste of mercury and calomel

Asbestos wool plug

Potassium chloride solution

Sintered glass disc

Figure 18. Calomel electrode

RELATIONSHIP BETWEEN POTENTIAL AND ACTIVITY

Equation (4.1), which gives the variation of the potential of an electrode with the activities of the constituents, may be deduced from the consideration of a cell in which a general redox electrode is combined with a standard hydrogen electrode. Consider the cell represented by the diagram

$$\text{Pt}; \text{H}_2(a = 1) \left| \text{H}^+(a = 1) \right| \left| \begin{array}{l} \text{oxidised form } (a = a_{\text{ox}}) \\ \text{reduced form } (a = a_{\text{red}}) \end{array} \right| \text{Pt}$$

Suppose that the e.m.f. of the cell is E_{cell} and that the redox electrode is its positive pole, the standard hydrogen electrode being the negative pole. If the cell is allowed to supply current, a reaction occurs in the cell, determined by its polarity. At the redox electrode, a reaction which consumes electrons will take place, and this may be represented as

$$p \text{ ox} + ze \rightarrow q \text{ red}$$

which indicates that p moles of the oxidised form combine with z moles of electrons to form q moles of the reduced form.

The reaction occurring at the hydrogen electrode will furnish electrons and must be

$$\tfrac{1}{2}H_2 \rightarrow H^+ + e$$

or

$$z/2\ H_2 \rightarrow zH^+ + ze$$

The overall cell reaction is the sum of the two electrode reactions, i.e.

$$p \text{ ox} + z/2\ H_2 \rightarrow q \text{ red} + zH^+$$

If the cell is operated in a thermodynamically reversible manner, the process occurs with maximum efficiency and the electrical work done is a maximum. Under conditions of constant temperature and pressure, the electrical work done will be equal to the decrease in the Gibbs free energy of the system.

For an amount of reaction corresponding to the stoichiometric quantities of reactants and products, i.e. for the reduction of an amount p of the oxidised form at activity a_{ox} to form an amount q of the reduced form at activity a_{red}, and the oxidation of an amount $z/2$ of hydrogen gas at unit activity to form an amount z of hydrogen ions at unit activity, z faradays of electricity must pass through the cell. In general, the quantity of electricity required is zF, where z = number of electrons involved in either electrode reaction, F = Faraday's constant. An amount of charge, zF, then, is passed through a potential difference of E_{cell}, and hence the electrical work done is zFE_{cell}. The work done *by* the system at constant temperature and pressure is equal to the *decrease* in the free energy, $-\Delta G$, for the cell reaction as written, hence

$$- \Delta G = zFE_{\text{cell}} \qquad (4.15)$$

If all the constituents of the cell had been in the standard state, the e.m.f. of the cell would have been the standard e.m.f. of the cell $E_{\text{cell}}°$. Furthermore, the decrease of free energy associated with the reaction would have been the standard decrease in free energy, $- \Delta G°$, whence

$$- \Delta G° = zFE_{\text{cell}}° \qquad (4.16)$$

The increase in free energy, ΔG, for a reaction is related to the standard increase in free energy, $\Delta G°$, by the van't Hoff reaction isotherm which states

$$\Delta G = \Delta G° + RT \ln \frac{\Pi(\text{activities of products})}{\Pi(\text{activities of reactants})} \qquad (4.17)$$

Applying eqn. (4.17) to the cell reaction, we have

$$\Delta G = \Delta G° + RT \ln \frac{a_{\text{red}}{}^q \times a_{\text{H}^+}{}^2}{a_{\text{ox}}{}^p \times a_{\text{H}_2}{}^{z/2}} \qquad (4.18)$$

As the hydrogen electrode is in the standard state, however, the activities of the hydrogen gas and the hydrogen ions are both unity, and eqn. (4.18) reduces to

$$\Delta G = \Delta G° + RT \ln \frac{a_{\text{red}}{}^q}{a_{\text{ox}}{}^p} \qquad (4.19)$$

Substituting in eqn. (4.19) from (4.15) and (4.16)

$$- zFE_{\text{cell}} = - zFE_{\text{cell}}° + RT \ln \frac{a_{\text{red}}{}^q}{a_{\text{ox}}{}^p}$$

or

$$E_{\text{cell}} = E_{\text{cell}}° - \frac{RT}{zF} \ln \frac{a_{\text{red}}{}^q}{a_{\text{ox}}{}^p} \qquad (4.20)$$

Equation (4.20) may be written in the alternative form

$$E_{\text{cell}} = E_{\text{cell}}° + \frac{RT}{zF} \ln \frac{a_{\text{ox}}{}^p}{a_{\text{red}}{}^q} \qquad (4.21)$$

The e.m.f. of the cell will be given by eqn. (4.10) as

$$E_{cell} = E_{redox} - E_{H_2}{}^\circ$$

or, with the cell in the standard state

$$E_{cell}{}^\circ = E_{redox}{}^\circ - E_{H_2}{}^\circ$$

Since $E_{H_2}{}^\circ$ is taken as zero by convention

$$E_{cell} = E_{redox} \quad \text{and} \quad E_{cell}{}^\circ = E_{redox}{}^\circ$$

and, substituting in eqn. (4.21), we have

$$E_{redox} = E_{redox}{}^\circ + \frac{RT}{zF} \ln \frac{a_{ox}{}^p}{a_{red}{}^q}$$

which is identical with eqn. (4.1).

THERMODYNAMICS OF CELLS

The relationship between the free energy change of a cell reaction and the e.m.f. of the cell is given by eqn. (4.15) which may be written

$$\Delta G = - zFE \tag{4.22}$$

where E is the e.m.f. of the cell.

The condition for equilibrium in a system at constant temperature and pressure is given by thermodynamics as $\Delta G = 0$. It can be understood, then, that if the components of an electrochemical cell were at their equilibrium activities, the e.m.f. of the cell would be zero.

The condition for a spontaneous change is that ΔG for that change should be negative, and the further removed from zero its value, the further from equilibrium is the system. It will be appreciated that, if the components of a cell are not at their equilibrium activities, ΔG for the spontaneous cell reaction will have a finite negative value and the e.m.f. of the cell will be finite. If the cell is allowed to operate spontaneously by supplying current to an external circuit, this electrical energy is supplied by the decrease in the free energy of the cell as the reaction proceeds.

It will continue until equilibrium is reached, by which time the e.m.f. of the cell will have decreased to zero.

Equation (4.22) provides a method for determining free energy changes of reactions. If a cell can be devised in which the particular reaction occurs, the free energy change can be calculated from the e.m.f. of the cell.

The enthalpy change for a reaction may also be deduced from e.m.f. measurements. The Gibbs–Helmholtz equation gives

$$\Delta H = \Delta G - T \left[\frac{\partial(\Delta G)}{\partial T} \right]_p$$

Substituting for ΔG from eqn. (4.22)

$$\Delta H = - zFE - T \left[\frac{\partial(-zFE)}{\partial T} \right]_p$$

As z is a constant for a given amount of reaction and F is a constant, we may write

$$\Delta H = - zFE + zFT \left(\frac{\partial E}{\partial T} \right)_p \tag{4.23}$$

From a measurement of the e.m.f. of the cell and the rate of change of e.m.f. with temperature, ΔH for the cell reaction may be calculated.

Entropy changes of cell reactions may also be determined from the temperature coefficient of the cell e.m.f.

$$\Delta S = - \left[\frac{\partial(\Delta G)}{\partial T} \right]_p$$

Substituting from eqn. (4.22)

$$\Delta S = - \left[\frac{\partial(-zFE)}{\partial T} \right]_p$$

Once again as z is constant for a given amount of reaction and F is a constant

$$\Delta S = - zF \left(\frac{\partial E}{\partial T} \right)_p \tag{4.24}$$

Although electrochemical methods of determining thermo-dynamic quantities are most attractive in theory, the practical difficulties of obtaining accurate results are very great. Unless immense precautions are taken during an experiment, the values of ΔG, ΔH and ΔS obtained from electrical measurements can only be regarded as approximate.

THERMODYNAMIC ASPECTS OF THE SIGN CONVENTION

Earlier in this Chapter the sign convention was introduced as a method of conveying the e.m.f. of a cell together with its polarity. For example, a Daniell cell with an observed e.m.f. of 1 V (positive pole Cu, negative pole Zn) could be represented by

$$Zn/ZnSO_4//CuSO_4/Cu \qquad E = +1 \text{ V} \qquad \text{(A)}$$

or

$$Cu/CuSO_4//ZnSO_4/Zn \qquad E = -1 \text{ V} \qquad \text{(B)}$$

Now that we have seen that the free energy change of a chemical reaction may be calculated from the e.m.f. of a cell the sign convention must be extended. In order that the expression $\Delta G = -zFE$ may be applied to the above notation the cell diagram must be considered to imply a certain cell reaction. The e.m.f. of cell A is positive and will therefore give a negative value of ΔG for the cell reaction. As a negative value of ΔG corresponds to a spontaneous reaction, the reaction implied by cell A must be the spontaneous reaction of the Daniell cell

$$Zn + Cu^2 \rightarrow Zn^{2+} + Cu \qquad \text{(A)}$$

Conversely, the e.m.f. of cell B is negative and will thus give a positive value of ΔG. As ΔG for cell B is equal and opposite to that for cell A, the cell B reaction must be the reverse of that of cell A.

$$Cu + Zn^{2+} \rightarrow Cu^{2+} + Zn \qquad \text{(B)}$$

The cell reaction for cell A is written, by convention, as reaction A and that for cell B as reaction B. Each cell diagram therefore implies a particular reaction which is always made up

113

of an oxidation reaction at the left-hand electrode and a reduction reaction at the right-hand electrode. If the e.m.f. of the cell is positive, the implied reaction is the spontaneous one. This is an alternative method of deducing the spontaneous cell reaction to the one given earlier in the chapter.

MEAN IONIC ACTIVITIES

Up to this point, many expressions containing terms in the activities of individual ions have been developed. In practice it is not possible to measure the activity of a single ionic species. Ions of one type can never occur alone in solution. They must always be accompanied by counter-ions carrying opposite charges, in order that the electrical neutrality of the solution be preserved. It is impossible to have a solution containing only sodium ions: there must always be an equivalent number of anions present.

Any property of a solution which is measured in the hope of gaining information about the activity of the ions cannot be influenced by only one type of ion. Any measurement made must be characteristic of a combination of ionic species, and the ionic activities of solutions which may be obtained experimentally must depend in some way upon all the types of ion present in the solution.

The quantity usually obtained from experimental observations is the *mean ionic activity* of the solution. The determination of this quantity from potentiometric measurements will be considered in Chapter 6.

Suppose that a molecule of an electrolyte dissociates in solution to provide ν_+ cations and ν_- anions:

$$(\text{electrolyte}) \rightleftharpoons \nu_+ (\text{cation}) + \nu_- (\text{anion})$$

Let the activities of the cations and anions be a_+ and a_-, respectively. Then the mean ionic activity of the solution, $a_\pm$, is related to the individual ionic activities by the definition

$$(a_\pm)^\nu = a_+^{\nu_+} \times a_-^{\nu_-} \tag{4.25}$$

where

$$\nu = \nu_+ + \nu_-$$

Although individual ionic activities cannot be measured, it is useful to retain the concept for the purposes of deriving the relationships between the electrode potentials of single electrodes and ionic activities. Potential measurements cannot be made on single electrodes, and when two electrodes are combined to form a complete cell, the expression obtained for its e.m.f. will contain ionic activity terms which can be related to a mean ionic activity if the problem is treated exactly.

CONCENTRATION CELLS

If two half cells of the same chemical nature are combined to form an electrochemical cell, an e.m.f. will arise if there is some difference in activity between them. This may be between the solutions or between the electrodes. Such cells are called *concentration cells*. We shall consider first some examples of those cases where the difference in activity lies between the solutions.

(i) *Identical electrodes in solutions of different activities*

   (a) Consider the cell represented by the diagram

$$Ag/AgNO_3//AgNO_3/Ag$$

$$(a_+)_1 \qquad (a_+)_2$$

The double solidus in the centre of the cell represents the elimination of the liquid junction potential, as would be very nearly the case if the two half cells were joined with a salt bridge. Suppose that the activities of the silver ions in the right- and left-hand electrodes are $(a_+)_2$ and $(a_+)_1$, respectively, the potentials of the electrodes being E_2 and E_1. Eqn. (4.5) gives

$$E_2 = E_{Ag}{}^\circ + \frac{RT}{F} \ln (a_+)_2$$

and

$$E_1 = E_{Ag}{}^\circ + \frac{RT}{F} \ln (a_+)_1$$

115

The e.m.f. of the cell will be given by eqn. (4.10) as

$$E_{cell} = E_2 - E_1$$

$$= \left(E_{Ag}^\circ + \frac{RT}{F}\ln (a_+)_2\right) - \left(E_{Ag}^\circ + \frac{RT}{F}\ln (a_+)_1\right)$$

$$= \frac{RT}{F}\ln (a_+)_2 - \frac{RT}{F}\ln (a_+)_1$$

$$= \frac{RT}{F}\ln \frac{(a_+)_2}{(a_+)_1} \tag{4.26}$$

If $(a_+)_2 > (a_+)_1$ eqn. (4.26) shows that the e.m.f. of the cell will be a positive quantity indicating that the right-hand electrode is the positive pole of the cell. Let us suppose this to be the case. If the cell is allowed to operate, an oxidation reaction occurs at the negative pole, i.e. the dissolution of metallic silver to form silver ions, at activity $(a_+)_1$

$$Ag \rightarrow Ag^+ + e$$

$$(a_+)_1$$

A corresponding reduction reaction at the positive pole will be the deposition of metallic silver from a solution of silver ions, at activity $(a_+)_2$

$$Ag^+ + e \rightarrow Ag$$

$$(a_+)_2$$

The overall cell reaction is thus

$$Ag^+ \rightarrow Ag^+$$

$$(a_+)_2 \quad (a_+)_1$$

the electrical neutrality of the solution being maintained by selective diffusion of cations and anions from the salt bridge into the electrode solutions.

As the reaction proceeds, the more concentrated solution becomes more dilute and the more dilute solution more concentrated, until finally both have the same activity. This corresponds to equilibrium when the e.m.f. of the cell is zero.

It has been mentioned above that the activity of an individual ionic species cannot be obtained. It is thus not possible to calculate the e.m.f. of the concentration cell from eqn. (4.26) as it stands, and assumptions have to be made.

Let the activities of the nitrate ions in the right- and left-hand electrodes be $(a_-)_2$ and $(a_-)_1$, respectively. The mean ionic activities of the silver nitrate solutions on the right and left, $(a_\pm)_2$ and $(a_\pm)_1$, respectively, are given, according to eqn. (4.25), by

$$(a_\pm)_2{}^2 = (a_+)_2(a_-)_2$$

and

$$(a_\pm)_1{}^2 = (a_+)_1 (a_-)_1$$

These relationships may be combined to give

$$\frac{(a_\pm)_2{}^2}{(a_\pm)_1{}^2} = \frac{(a_+)_2(a_-)_2}{(a_+)_1(a_-)_1}$$

If it is now assumed that the ratio of the activities of the nitrate ions is equal to that of the silver ions, we have $(a_+)_2/(a_+)_1 = (a_-)_2/(a_-)_1$, and hence

$$\frac{(a_\pm)_2{}^2}{(a_\pm)_1{}^2} = \frac{(a_+)_2{}^2}{(a_+)_1{}^2}$$

Equation (4.26) may now be written as

$$E_{\text{cell}} = \frac{RT}{F} \ln \frac{(a_\pm)_2}{(a_\pm)_1} \tag{4.27}$$

In general, we have for this type of concentration cell where z electrons are involved in the electrode reactions

$$E_{\text{cell}} = \frac{RT}{zF} \ln \frac{(a_\pm)_2}{(a_\pm)_1} \tag{4.28}$$

(b) If we now consider the same cell where the liquid junction has not been eliminated, the result is somewhat different:

117

$$Ag/AgNO_3 \mid AgNO_3/Ag$$

$$(a_+)_1, (a_-)_1 \mid (a_+)_2, (a_-)_2$$

Using the same notation as before, the right-hand electrode will once again be the positive pole of the cell, the left-hand one the negative. We have seen already that the reaction at the right-hand electrode is

$$Ag^+ + e \to Ag$$

$$(a_+)_2$$

Electrons are thus entering the solution at the right-hand electrode, and according to the definitions given in Chapter 1, this must be the *cathode*. Conversely, the left-hand electrode is the *anode*. It will be noticed that, in this case, the cathode is positive and the anode is negative, which is the reverse of the situation which occurs in electrolysis. It is to be remembered that when energy is supplied *to* the system from an external source (electrolysis), anodes are positive and cathodes are negative. When energy is supplied *by* the system (electrochemical cell), anodes are negative and cathodes are positive.

Consider now the passage of 96 487 C of electricity through the cell. At the left-hand electrode, one mol of silver will dissolve to form silver ions. At the same time, t_+ mol of silver ion will migrate across the liquid junction towards the right-hand electrode (cathode), where t_+ is the average transport number of the silver ion for the two concentrations involved in the liquid junction. The net gain of silver ions in the left-hand electrode is thus $(1 - t_+)$ mol. Similarly, t_- mol of nitrate ion will migrate across the liquid junction towards the left-hand electrode (anode). The change in the solution may be stated as

gain of silver ions at activity $(a_+)_1 = (1 - t_+)$ mol $= t_-$ mol

gain of nitrate ions at activity $(a_-)_1 = t_-$ mol

At the right-hand electrode, one mol of silver ions will have been deposited but, at the same time, t_+ mol will have migrated into the solution across the liquid junction. The net loss of silver ions is thus $(1 - t_+)$ or t_- mol. Also t_- mol of nitrate ion will have migrated away from this electrode across the liquid junction. The change in the solution of the right-hand electrode may be stated

loss of silver ions at activity $(a_+)_2 = t_-$ mol

loss of nitrate ions at activity $(a_-)_2 = t_-$ mol

The overall cell reaction for 96 487 C of electricity is thus

$$t_-Ag^+ + t_-NO_3^- \rightarrow t_-Ag^+ + t_-NO_3^-$$
$$(a_+)_2 \quad (a_-)_2 \quad (a_+)_1 \quad (a_-)_1$$

Applying the van't Hoff reaction isotherm, eqn. (4.17)

$$\Delta G = \Delta G^\circ + RT \ln \frac{(a_+)_1{}^{t-} \times (a_-)_1{}^{t-}}{(a_+)_2{}^{t-} \times (a_-)_2{}^{t-}} \tag{4.29}$$

For silver nitrate, a uni-univalent electrolyte, eqn. (4.25) gives

$$a_+ \times a_- = a_\pm{}^2$$

and hence

$$a_+{}^{t-}a_-{}^{t-} = a_\pm{}^{2t-}$$

Substituting in eqn. (4.29)

$$\Delta G = \Delta G^\circ + 2t_-RT \ln \frac{(a_\pm)_1}{(a_\pm)_2} \tag{4.30}$$

Now ΔG° is the change in free energy when all the reactants and products are in their standard states of unit activity. If this were so, the products would be identical with the reactants and hence the free energy of the products would be exactly equal to that of the reactants. ΔG° must therefore be zero. Equation (4.30) thus reduces to

$$\Delta G = 2t_-RT \ln \frac{(a_\pm)_1}{(a_\pm)_2}$$

Making the substitution $\Delta G = -zFE_{cell}$ where $z = 1$

119

$$- FE_{\text{cell}} = 2t_- RT \ln \frac{(a_\pm)_1}{(a_\pm)_2}$$

or

$$E_{\text{cell}} = 2t_- \frac{RT}{F} \ln \frac{(a_\pm)_2}{(a_\pm)_1} \qquad (4.31)$$

Equation (4.31) gives the complete e.m.f. of the cell with a liquid junction. It will be noticed that the transport number occurring in the equation is that of the anion. In the cell considered, the electrodes are in equilibrium with the cations in the solutions, and it always happens that the transport number which occurs in equations of the type of (4.31) is that of the ion to which the electrodes are *not* reversible.

To arrive at the value of the liquid junction potential, we have only to subtract eqn. (4.27) from (4.31), when we have

$$E_l = 2t_- \frac{RT}{F} \ln \frac{(a_\pm)_2}{(a_\pm)_1} - \frac{RT}{F} \ln \frac{(a_\pm)_2}{(a_\pm)_1}$$

or

$$E_l = (2t_- - 1) \frac{RT}{F} \ln \frac{(a_\pm)_2}{(a_\pm)_1} \qquad (4.32)$$

where E_l is the liquid junction potential. Remembering that $t_+ + t_- = 1$, eqn. (4.32) may be written in the form

$$E_l = (t_- - t_+) \frac{RT}{F} \ln \frac{(a_\pm)_2}{(a_\pm)_1} \qquad (4.33)$$

which is identical with eqn. (4.8).

It must be remembered that eqn. (4.32) and (4.33) are not exact, owing to the assumption that was made in deriving eqn. (4.27).

It can be seen from eqn. (4.33) that the liquid junction potential may reinforce or oppose the e.m.f., due to the electrode potentials, according to whether t_- or t_+ is greater. If the transport number of the anion is greater than that of the cation, the anion will diffuse more rapidly across the boundary. As both ions diffuse

120

from the more concentrated to the more dilute solution, there would be a net transfer of negative charge across the boundary from right to left. The separation of charge across the boundary in this way reinforces the e.m.f. due to the electrodes, as the left-hand one is already the negative pole of the cell in view of the assumption that $(a_+)_2 > (a_+)_1$

The general equation for the e.m.f. of a concentration cell with a liquid junction where the electrodes are in equilibrium with positive ions, is

$$E_{cell} = t_- \cdot \frac{\nu}{\nu_+} \frac{RT}{zF} \ln \frac{(a_\pm)_2}{(a_\pm)_1} \qquad (4.34)$$

The general equation for the liquid junction potential in such a cell is

$$E_l = \left(t_- \cdot \frac{\nu}{\nu_+} - 1 \right) \frac{RT}{zF} \ln \frac{(a_\pm)_2}{(a_\pm)_1} \qquad (4.35)$$

(c) A concentration cell may also be constructed by combining two electrodes of the second kind, for example

Ag/AgCl(s), KCl//KCl, AgCl(s)/Ag

$$(a_-)_2 \qquad (a_-)_1$$

Suppose for simplicity that the liquid junction potential has been eliminated with a salt bridge and that the activities of the chloride ions in the left- and right-hand electrodes are $(a_-)_2$ and $(a_-)_1$, respectively. Suppose also that the potentials of the two electrodes are E_2 and E_1.

Applying eqn. (4.13), we have

$$E_2 = E_{AgCl}^\circ - \frac{RT}{F} \ln (a_-)_2$$

$$E_1 = E_{AgCl}^\circ - \frac{RT}{F} \ln (a_-)_1$$

From eqn. (4.10), the e.m.f. of the cell, E_{cell}, will be given by

$$E_{cell} = E_1 - E_2$$

$$= \left(E_{AgCl}° - \frac{RT}{F} \ln (a_-)_1 \right)$$

$$- \left(E_{AgCl}° - \frac{RT}{F} \ln (a_-)_2 \right)$$

$$= \frac{RT}{F} \ln (a_-)_2 - \frac{RT}{F} \ln (a_-)_1$$

$$= \frac{RT}{F} \ln \frac{(a_-)_2}{(a_-)_1} \qquad (4.36)$$

Once again, eqn. (4.36) cannot be used directly to calculate E_{cell}, and the approximation has to be made that the ratio of the activities of the chloride ions in the two solutions is equal to that of the potassium ions. The ratio $(a_-)_2/(a_-)_1$ may now be put equal to $(a_\pm)_2/(a_\pm)_1$, and eqn. (4.36) may be written

$$E_{cell} = \frac{RT}{F} \ln \frac{(a_\pm)_2}{(a_\pm)_1} \qquad (4.37)$$

If $(a_\pm)_2 > (a_\pm)_1$, eqn. (4.37) shows that the e.m.f. of the cell will be a positive quantity indicating that the right-hand electrode is the positive pole of the cell.

If the liquid junction had not been eliminated, the e.m.f. of the cell would have been given exactly by

$$E_{cell} = 2t_+ \frac{RT}{F} \ln \frac{(a_\pm)_2}{(a_\pm)_1} \qquad (4.38)$$

The transport number in eqn. (4.38) is that of the cation, as the electrodes are in equilibrium with the anions in solution. The contribution of the liquid junction potential is thus given approximately by

$$E_l = (t_+ - t_-) \frac{RT}{F} \ln \frac{(a_\pm)_2}{(a_\pm)_1} \qquad (4.39)$$

The general equation for the e.m.f. of a concentration cell where the electrodes are in equilibrium with negative ions and the half cells are joined by a salt bridge, is

$$E_{cell} = \frac{RT}{zF} \ln \frac{(a_{\pm})_2}{(a_{\pm})_1} \tag{4.40}$$

If the two half cells join at a liquid junction, the general expression for the e.m.f. is

$$E_{cell} = t_+ \frac{\nu}{\nu_-} \frac{RT}{zF} \ln \frac{(a_{\pm})_2}{(a_{\pm})_1} \tag{4.41}$$

and the general equation for the liquid junction potential

$$E_l = \left(t_+ \cdot \frac{\nu}{\nu_-} - 1 \right) \frac{RT}{zF} \ln \frac{(a_{\pm})_2}{(a_{\pm})_1} \tag{4.42}$$

The types of concentration cells discussed in Sections (a), (b) and (c) are sometimes called *concentration cells with transport*, as there is direct transport of ions across the liquid junctions. Although salt bridges minimise liquid junction potentials, it is possible to devise a concentration cell without any liquid junction, thus avoiding the nuisance of liquid junction potentials. Such a cell is sometimes called a *concentration cell without transport*.

(d) An example of a concentration cell without transport is

Zn/ ZnSO$_4$,   Hg$_2$SO$_4$(s)/Hg/Hg$_2$SO$_4$(s),   ZnSO$_4$ /Zn

$(a_+)_1, (a_-)_1$ ⠀⠀⠀⠀⠀⠀⠀⠀⠀⠀⠀⠀⠀⠀⠀⠀$(a_+)_2, (a_-)_2$

In this cell the activities of the zinc and sulphate ions differ. Suppose that they are $(a_+)_1$ and $(a_-)_1$ on the left and $(a_+)_2$ and $(a_-)_2$ on the right. The mean ionic activities of the zinc sulphate solutions on the left and right, $(a_{\pm})_1$ and $(a_{\pm})_2$, respectively, will be given by

$$(a_{\pm})_1{}^2 = (a_+)_1(a_-)_1$$

and

$$(a_{\pm})_2{}^2 = (a_+)_2(a_-)_2$$

123

The cell is really a double cell consisting of two single cells connected in series, and might alternatively be represented as

$$Zn/\ ZnSO_4,\ \ Hg_2SO_4(s)/Hg\text{------}Hg/Hg_2SO_4(s),\ \ ZnSO_4/Zn$$

$$(a_+)_1, (a_-)_1 \hspace{6cm} (a_+)_2, (a_-)_2$$

It is seen that each single cell is a chemical cell consisting of a simple zinc electrode and a mercury–mercurous sulphate electrode. In order to arrive at the e.m.f. of the double cell, it will be best to derive the expressions for the e.m.f. of the individual single cells.

Suppose that the e.m.f. of the cell containing the zinc sulphate at mean ionic activity, $(a_\pm)_1$, is E^I. From eqn. (4.10) the e.m.f. of the cell will be given by

$$E_I = E_{Hg_2SO_4} - E_{Zn}$$

The potential of the zinc electrode is given by eqn. (4.5)

$$E_{Zn} = E_{Zn}^\circ + \frac{RT}{2F} \ln (a_+)_1$$

and the potential of the mercurous sulphate electrode, which is an electrode of the second kind and is thus in equilibrium with the anions, will be given by eqn. (4.7)

$$E_{Hg_2SO_4} = E_{Hg_2SO_4}^\circ - \frac{RT}{2F} \ln (a_-)_1$$

The e.m.f. of the cell is thus

$$E^I = \left(E_{Hg_2SO_4}^\circ - \frac{RT}{2F} \ln (a_-)_1 \right) - \left(E_{Zn}^\circ + \frac{RT}{2F} \ln (a_+)_1 \right)$$

$$= (E_{Hg_2SO_4}^\circ - E_{Zn}^\circ) - \frac{RT}{2F} \ln (a_+)_1 (a_-)_1$$

$$= (E_{Hg_2SO_4}^\circ - E_{Zn}^\circ) - \frac{RT}{2F} \ln (a_\pm)_1{}^2$$

$$= (E_{Hg_2SO_4}^\circ - E_{Zn}^\circ) - \frac{RT}{F} \ln (a_\pm)_1 \hspace{2cm} (4.43)$$

As the right-hand cell is the same as the left-hand one in all respects except the activity of the zinc sulphate solution and the reversal of the electrodes, the e.m.f. of the right-hand cell, denoted E^{II}, will be given by an equation similar to eqn. (4.43), in which $(a_\pm)_2$ is substituted for $(a_\pm)_1$ and all the signs are reversed

$$E^{II} = E_{Zn}^\circ - E_{Hg_2SO_4}^\circ) + \frac{RT}{F} \ln (a_\pm)_2 \qquad (4.44)$$

In the double cell, the two single cells are connected in series, so that the e.m.f. of the complete double cell, E_{cell}, will be equal to the sum of the e.m.f. of the single cells.
Thus

$$E_{cell} = E^I + E^{II}$$

$$= \frac{RT}{F} \ln (a_\pm)_2 - \frac{RT}{F} \ln (a_\pm)_1$$

$$= \frac{RT}{F} \ln \frac{(a_\pm)_2}{(a_\pm)_1} \qquad (4.45)$$

If $(a_\pm)_2 > (a_\pm)$, E_{cell} will be positive indicating that the zinc electrode on the right is the positive pole of the complete cell.

The general expressions for the e.m.f. of a concentration cell without transport are

$$E_{cell} = \frac{\nu}{\nu_+} \cdot \frac{RT}{zF} \ln \frac{(a_\pm)_2}{(a_\pm)_1} \qquad (4.46)$$

where the outer electrodes are in equilibrium with positive ions, and

$$E_{cell} = \frac{\nu}{\nu_-} \cdot \frac{RT}{zF} \ln \frac{(a_\pm)_2}{(a_\pm)_1} \qquad (4.47)$$

where the outer electrodes are in equilibrium with negative ions.

(ii) Electrodes of different activities in the same solution

(e) An example of this type of concentration cell is provided by a cell which has two gas electrodes at different pressures

$$\text{Pt; Cl}_2/ \quad \text{HCl} \quad /\text{Cl}_2\text{; Pt}$$

$$p_1 \qquad\qquad\qquad p_2$$

Taking the potential of the electrode on the left as E_1, with the chlorine gas at a pressure of p_1, and that of the right-hand one, as E_2, with the chlorine at a pressure of p_2, suppose that the activity of the chloride ions in solution is a_-.

The equilibrium at each electrode may be represented

$$\tfrac{1}{2}\text{Cl}_2 + e \rightleftharpoons \text{Cl}^-$$

Assuming that the activity of the gas is equal to its pressure and applying eqn. (4.1) to the left-hand electrode

$$E_1 = E_{\text{Cl}_2}{}^\circ + \frac{RT}{F} \ln \frac{p_1{}^{1/2}}{a_-}$$

or

$$E_1 = E_{\text{Cl}_2}{}^\circ + \frac{RT}{F} \ln p_1{}^{1/2} - \frac{RT}{F} \ln a_- \tag{4.48}$$

Similarly, for the right-hand electrode we have

$$E_2 = E_{\text{Cl}_2}{}^\circ + \frac{RF}{F} \ln p_2{}^{1/2} - \frac{RT}{F} \ln a_- \tag{4.49}$$

Equation (4.10) shows that the e.m.f. of the cell is given by

$$E_{\text{cell}} = E_2 - E_1$$

$$= \frac{RT}{F} \ln p_2{}^{1/2} - \frac{RT}{F} \ln p_1{}^{1/2}$$

$$= \frac{RT}{F} \ln \left(\frac{p_2}{p_1} \right)^{1/2}$$

$$= \frac{RT}{2F} \ln \frac{p_2}{p_1} \tag{4.50}$$

As before, with $p_2 > p_1$ the electrode on the right will be the positive pole of the cell.

(f) A further example of a concentration cell with electrodes of different activities is provided by a cell which has amalgam electrodes, such as

$$Zn/Hg \mid ZnSO_4 \mid Zn/Hg$$

$$a_2 \qquad\qquad a_1$$

Suppose that the activity of the zinc in the left-hand amalgam electrode is a_2, in the right-hand one, a_1, and that the activity of the zinc ions in the solution is a_+. The equilibrium at each electrode will be

$$Zn \rightleftharpoons Zn^{2+} + 2e$$

If the potential of the left-hand electrode is E_2 and that of the right-hand one E_1, eqn. (4.1) gives

$$E_2 = E_{Zn}^{\bullet} + \frac{RT}{2F} \ln \frac{a_+}{a_2}$$

$$= E_{Zn}^{\bullet} + \frac{RT}{2F} \ln a_+ - \frac{RT}{2F} \ln a_2$$

and similarly

$$E_1 = E_{Zn}^{\bullet} + \frac{RT}{2F} \ln a_+ - \frac{RT}{2F} \ln a_1$$

From eqn. (4.10), the e.m.f. of the cell, E_{cell}, is given by

$$E_{cell} = E_1 - E_2$$

$$= -\frac{RT}{2F} \ln a_1 - \left(-\frac{RT}{2F} \ln a_2 \right)$$

$$= \frac{RT}{2F} \ln \frac{a_2}{a_1} \qquad\qquad (4.51)$$

With $a_2 > a_1$ the right-hand electrode will be the positive pole of the cell.

REDOX SYSTEMS

Although there is no fundamental difference between electrode and redox potentials, the latter are frequently considered separately, as they have greater application in the field of oxidation–reduction reactions in solution. *Table 14* gives some values of standard redox potentials for various redox systems.

It will be remembered that a redox potential is the potential adopted by an inert electrode due to the equilibrium between the oxidised and reduced forms of a system in solution. It may be represented

$$ox + e \rightleftharpoons red$$

If the equilibrium lies to the left, the electrons which are generated collect on the electrode, giving rise to a negative potential. This means that the redox system readily gives up electrons, which is tantamount to saying that it is a good reducing agent. Alterna-

Table 14

STANDARD REDOX POTENTIALS AT 25°C

	V
Co^{3+}/Co^{2+}	+ 1·82
Ce^{4+}/Ce^{3+}	+ 1·61
MnO_4^-/Mn^{2+}	+ 1·51
$Cr_2O_7^{2-}/Cr^{3+}$	+ 1·33
Fe^{3+}/Fe^{2+}	+ 0·771
$Fe(CN)_6^{3-}/Fe(CN)_6^{4-}$	+ 0·36
Cu^{2+}/Cu^+	+ 0·153
Sn^{4+}/Sn^{2+}	+ 0·15
$Co(NH_3)_6^{3+}/Co(NH_3)_6^{2+}$	+ 0·10
Cr^{3+}/Cr^{2+}	− 0·41

(After W. M. LATIMER *Oxidation Potentials*, 2nd ed., New York, Prentice-Hall, 1952)

tively, if the equilibrium lies to the right, electrons are abstracted from the inert electrode which thus adopts a positive potential. In this case, the redox system readily accepts electrons and is thus an oxidising agent.

It is true to say that the more positive the redox potential of a system, the better oxidising agent is that system. It follows that a system with a more positive potential will oxidise one with a less positive potential or, alternatively, that a system with a more negative potential will reduce one with a less negative potential.

If the cerous–ceric system in its standard state were mixed with the ferrous–ferric system in its standard state, it may be seen from *Table 14* that the ceric ions would oxidise the ferrous ions, as the cerium system has a more positive potential than the iron system. Similarly, if the ferrous–ferric system were mixed with the stannous–stannic, the stannous ions would reduce the ferric ions.

In passing, it is interesting to note the effect of complexing on the redox potential of a system. This is shown in *Table 14* by the cyanide complexes of iron and the ammine complexes of cobalt.

STANDARD POTENTIALS AND EQUILIBRIUM CONSTANTS

We must now consider more closely the effects of mixing two redox systems. Suppose we have a solution containing ferrous and ferric ions and another containing stannous and stannic ions. The potential of the iron system, E_{Fe}, and that of the tin system, E_{Sn}, will be given by eqn. (4.1) as

$$E_{Fe} = E_{Fe}^{\circ} + \frac{RT}{F} \ln \frac{a_{Fe^{3+}}}{a_{Fe^{2+}}} \qquad (4.52)$$

and

$$E_{Sn} = E_{Sn}^{\circ} + \frac{RT}{2F} \ln \frac{a_{Sn^{4+}}}{a_{Sn^{2+}}} \qquad (4.53)$$

If we now mix these two solutions, the tin system will reduce the iron system and the reaction will be

$$Sn^{2+} + 2Fe^{3+} \rightarrow 2Fe^{2+} + Sn^{4+}$$

129

This reaction occurs because the potential of the tin system is lower than that of the iron system. As the reaction proceeds, the activity of the ferric ions decreases and that of the ferrous ions increases. From eqn. (4.52) it will be seen that both these changes will result in a lowering of E_{Fe}. Simultaneously, the activity of the stannic ions increases and that of the stannous ions decreases. Equation (4.53) shows that these changes result in an increase of E_{Sn}. As the reaction proceeds, the potential of the iron system falls towards that of the tin system which is simultaneously increasing. At some point, the potentials of the two systems become equal, and at this stage the reaction stops, indicating that equilibrium has been reached. At equilibrium, then

$$E_{Fe} = E_{Sn}$$

and substituting from eqn. (4.52) and (4.53) we may write

$$E_{Fe}^{\circ} + \frac{RT}{F} \ln \frac{a_{Fe^{3+}}}{a_{Fe^{2+}}} = E_{Sn}^{\circ} + \frac{RT}{2F} \ln \frac{a_{Sn^{4+}}}{a_{Sn^{2+}}}$$

Rearranging

$$\frac{RT}{2F} \ln \frac{a_{Sn^{4+}}}{a_{Sn^{2+}}} - \frac{RT}{F} \ln \frac{a_{Fe^{3+}}}{a_{Fe^{2+}}} = E_{Fe}^{\circ} - E_{Sn}^{\circ}$$

$$\frac{RT}{2F} \ln \frac{a_{Sn^{4+}}}{a_{Sn^{2+}}} - \frac{RT}{2F} \ln \frac{a_{Fe^{3+}}^{2}}{a_{Fe^{2+}}^{2}} = E_{Fe}^{\circ} - E_{Sn}^{\circ}$$

$$\frac{RT}{2F} \ln \frac{a_{Sn^{4+}} \times a_{Fe^{2+}}^{2}}{a_{Sn^{2+}} \times a_{Fe^{3+}}^{2}} = E_{Fe}^{\circ} - E_{Sn}^{\circ}$$

It will be observed that the logarithmic term is the equilibrium constant of the reaction, and putting this equal to K, we have

$$\frac{RT}{2F} \ln K = E_{Fe}^{\circ} - E_{Sn}^{\circ}$$

$$\ln K = (E_{Fe}^{\circ} - E_{Sn}^{\circ}) \frac{2F}{RT} \qquad (4.54)$$

If the reaction had been carried out in an electrochemical cell consisting of a ferrous–ferric electrode and a stannous–stannic electrode, to imply the above reaction the iron electrode would have to be written on the right and the standard e.m.f. of the cell would have been $(E_{Fe}^{\circ} - E_{Sn}^{\circ})$. Equation (4.53) thus leads to the general result

$$\ln K = \frac{zFE_{cell}^{\circ}}{RT} \qquad (4.55)$$

Although eqn. (4.55) was deduced from a consideration of the interaction between two redox systems, it is a perfectly general result which applies to any cell reaction, irrespective of whether the cell consists of redox or normal electrode systems.

The same result may be achieved by combining eqn. (4.16)

$$\Delta G^{\circ} = - zFE_{cell}^{\circ}$$

with the thermodynamic relationship

$$\Delta G^{\circ} = - RT \ln K$$

whence

$$- RT \ln K = - zFE_{cell}^{\circ}$$

$$\ln K = \frac{zFE_{cell}^{\circ}}{RT}$$

Standard redox or standard electrode potentials may thus be used to calculate equilibrium constants.

REDOX TITRATIONS

In redox titrations, redox systems are used as volumetric reagents for quantitative analysis. There are, however, restrictions upon the choice of systems which can be used. If, for example, a solution of ferrous ions is to be titrated with ceric ions, the reaction will be

$$Fe^{2+} + Ce^{4+} \rightarrow Fe^{3+} + Ce^{3+}$$

131

If the volume of ceric solution added is to be a measure of the amount of ferrous present in the original solution, it is necessary for the reaction to go to completion within the limits of accuracy of the titration. It is sufficient if the reaction goes to 99.9 per cent completion. This means that the amount of ferrous ion remaining is 0.1 per cent of the total iron content of the solution.

The equilibrium constant of the above reaction is given by

$$K = \frac{a_{Fe^{3+}} \times a_{Ce^{3+}}}{a_{Fe^{2+}} \times a_{Ce^{4+}}}$$

If the reaction is 99.9 per cent complete, the ratio of the activities of the ferric and ferrous ions will be given by

$$\frac{a_{Fe^{3+}}}{a_{Fe^{2+}}} = \frac{99.9}{0.1} \simeq 10^3$$

Similarly, the ratio of the cerous and ceric ion activities is

$$\frac{a_{Ce^{3+}}}{a_{Ce^{4+}}} = \frac{99.9}{0.1} \simeq 10^3$$

The equilibrium for the reaction is thus 10^6. If ceric ions are to oxidise ferrous ions quantitatively, the equilibrium constant of the reaction must not be less than 10^6.

The reaction could be carried out in an electrochemical cell such as

$$Pt/Fe^{3+}, Fe^{2+}//Ce^{3+}, Ce^{4+}/Pt$$

and its standard e.m.f. would be given by eqn. (4.10) as

$$E_{cell}^{\circ} = E_{Ce}^{\circ} - E_{Fe}^{\circ}$$

From *Table 14* it may thus be seen that

$$E_{cell}^{\circ} = 1.61 - 0.77 = 0.84 \text{ V.}$$

Substituting $z = 1$, $R = 8.314 \, JK^{-1} \, mol^{-1}$, $T = 298 \, K$, $F = 96\,500 \, C \, mol^{-1}$ into eqn. (4.55) and converting to common logarithms

$$\log K = \frac{0 \cdot 84}{0 \cdot 059} = 14 \cdot 23$$

$$K = 1 \cdot 7 \times 10^{14}$$

As this value is in excess of 10^6, it is apparent that ferrous ions could be quantitatively titrated with ceric ions.

If we now consider the limiting value of an equilibrium constant to be 10^6, that of the standard e.m.f. of the cell in which the reaction would occur at 25°C is given by

$$\log 10^6 = \frac{E_{\text{cell}}°}{0 \cdot 059 \text{V}}$$

$$E_{\text{cell}}° = 0 \cdot 059 \times 6 = 0 \cdot 35 \text{ V}$$

This means that, if two redox systems are to be suitable for redox titrations, their standard redox potentials must differ by at least $0 \cdot 35$V. This is only true for the case where $z = 1$ for both systems. If $z = 1$ for one and $z = 2$ for the other, it can be shown that the minimum value of K for quantitative conversion is 10^9 which, in turn, means that the standard redox potentials of the two systems must differ by at least $0 \cdot 26$ V.

We must now examine the change of potential which occurs during the course of a redox titration. Consider, for example, that of a solution of ferrous sulphate with a solution of ceric sulphate. After each addition of ceric sulphate, equilibrium will be reached and at all stages of the titration E_{Fe} will be equal to E_{Ce}. The potential at any stage of the titration may thus be calculated from either of the relationships

$$E_{\text{Fe}} = E_{\text{Fe}}° + \frac{RT}{F} \ln \frac{a_{\text{Fe}^{3+}}}{a_{\text{Fe}^{2+}}} \tag{4.56}$$

or

$$E_{\text{Ce}} = E_{\text{Ce}}° + \frac{RT}{F} \ln \frac{a_{\text{Ce}^{4+}}}{a_{\text{Ce}^{3+}}} \tag{4.57}$$

It is convenient to use eqn. (4.56) to calculate the potential before the end-point is reached and (4.57) after it has been passed. The potential at the end-point may be calculated in the following way.

As pointed out previously, the equilibrium constant of the reaction may be written

$$K = \frac{a_{Fe^{3+}} \times a_{Ce^{3+}}}{a_{Fe^{2+}} \times a_{Ce^{4+}}}$$

and this relationship will hold throughout the titration. At the end-point, however

$$\frac{a_{Fe^{3+}}}{a_{Fe^{2+}}} = \frac{a_{Ce^{3+}}}{a_{Ce^{4+}}}$$

Hence

$$K = \left(\frac{a_{Fe^{3+}}}{a_{Fe^{2+}}}\right)^2$$

or

$$\frac{a_{Fe^{3+}}}{a_{Fe^{2+}}} = \sqrt{K}$$

Substituting in eqn. (4.56), we have

$$E_e = E_{Fe}^{\circ} + \frac{RT}{F} \ln \sqrt{K}$$

where E_e is the potential at the end-point.

The variation of potential in the neighbourhood of the end-point for the titration of ferrous with ceric ions has been calculated by this method, and the results are plotted in *Figure 19*.

It must be pointed out that *Figure 19* represents a somewhat idealised situation. In actual practice, the potentials will be modified due to some complexing of the ferric and ceric ions with the sulphate ions which are present. The variation of the actual potential, however, will be of the same form and will show a rapid change near the end-point.

REDOX INDICATORS

The end-point of a redox titration can be detected by the use of a redox indicator. Such indicators are themselves redox systems, the oxidised and reduced forms of which have different colours.

It has been mentioned already that when two redox systems of different potentials are mixed, a reaction occurs in which the system with the higher potential oxidises that with the lower. This results in the approach of the two potentials, the higher one decreasing and the lower one increasing until they are equal, at which point the reaction has reached equilibrium. This point must now be examined in greater detail to ascertain the effect

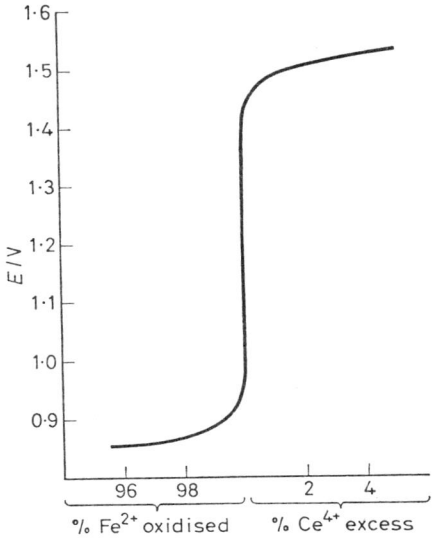

Figure 19. Variation of potential in titration of ferrous with ceric ions

of the relative amounts of the two redox systems on the changes of potential which occur.

Suppose that a small amount of a redox indicator is added to a relatively large amount of a solution of ferrous and ferric ions. Suppose also that the potential of the indicator is greater than that of the ferrous–ferric system. Ferrous ions will be oxidised by the indicator. Representing the oxidised form of the indicator as O and the reduced form as R, the reaction may be written

$$Fe^{2+} + O \rightarrow Fe^{3+} + R$$

assuming that there is a difference of only one electron between O and R.

As there is only a small amount of indicator present, it could be completely converted to the reduced form for the loss of a small amount of ferrous ions. If so, its potential will fall considerably, as there will have been an enormous change in the activity ratio a_O/a_R. As this is accomplished by only a small amount of ferrous ions in a relatively large amount of the ferrous–ferric system, the change in the ratio of the activities of the ferric to the ferrous ions will be very small indeed. The potential of the ferrous–ferric system will have increased only by an infinitesimal amount.

It can thus be understood that when a small amount of indicator is added to a relatively large amount of a redox system, the indicator adopts the potential of the system to which it is added. In the titration of ferrous sulphate with ceric sulphate, for example, the potential of any indicator present would follow the curve shown in *Figure 19*.

It has already been pointed out in Chapter 3 that the eye only detects a change of colour in an indicator between the limits of about 10 and 90 per cent mixtures of the differently coloured forms. If a redox indicator is in the fully reduced form and is gradually oxidised, the eye will detect a change in colour only when about 10 per cent of the indicator is in the oxidised form. The potential of the system at this point will be given by

$$E = E_{In}^{\circ} + \frac{RT}{zF} \ln \frac{0 \cdot 09}{0 \cdot 91}$$

$$= E_{In}^{\circ} + \frac{RT}{zF} \ln \frac{1}{10} \tag{4.58}$$

After 90 per cent of the indicator is in the oxidised form, the eye will detect no further colour change, and the potential at this point will be given by

$$E = E_{In}° + \frac{RT}{zF} \ln \frac{0·91}{0·09}$$

$$= E_{In}° + \frac{RT}{zF} \ln 10 \tag{4.59}$$

The range of potential over which the eye observes a colour change in a redox indicator is thus given by eqn. (4.58) and (4.59) in which $E_{In}°$ is the standard redox potential of the indicator. Assuming that $z = 1$ and that the temperature is 25°C the limits are from $E_{In}° - 0·059$ to $E_{In}° + 0·059$. The working range of a redox indicator where $z = 1$ may thus be taken as $E_{In}° \pm 0·06$ V, i.e. about 0·12 V.

For a redox titration to be quantitative, we already have the condition that the standard redox potentials of the two systems must differ by about 0·3 V. If this is the case, the rapid change of potential at the end-point should cover a range of at least 0·12 V. It should thus be possible to achieve a very sharp end-point with a redox indicator, provided that its working range is covered by the change of potential at the end-point of the titration. The ideal choice of indicator is obviously one with a standard redox potential which is equal to the end-point potential of the titration system.

5

APPLICATIONS OF CONDUCTANCE MEASUREMENTS

DETERMINATION OF LIMITING MOLAR CONDUCTIVITY

(*a*) *Strong electrolytes*—The limiting molar conductivity of strong electrolytes may be determined by an extrapolation method. The conductance is determined at various low concentrations and a graph is plotted of molar conductivity against the square root of concentration. This plot should give a straight line, as illustrated in *Figure 4* (p. 20) and it is easily extrapolated to obtain the limiting value of the molar conductivity.

(*b*) *Weak electrolytes*—In the case of weak electrolytes, limiting molar conductivity cannot be directly determined. If those of the constituent ions of the electrolyte at infinite dilution are known, that of the electrolyte may be obtained from Kohlrausch's law, eqn. (2.16)

$$\Lambda^\infty = v_+ \Lambda_+^\infty + v_- \Lambda_-^\infty$$

An example of this type of calculation has been given in Chapter 2.

Alternatively, the limiting molar conductivity of a weak electrolyte may be obtained by combining the values of this quantity for certain strong electrolytes. For example

$$\Lambda^\infty(CH_3COOH) = \Lambda^\infty(HCl) + \Lambda^\infty(CH_3COONa) - \Lambda^\infty(NaCl)$$

$$(5.1)$$

The validity of this relationship may be seen by splitting the terms on the right-hand side of the equation into their various ionic components:

$$\Lambda^\infty(\text{HCl}) + \Lambda^\infty(\text{CH}_3\text{COONa}) - \Lambda^\infty(\text{NaCl})$$

$$= \Lambda^\infty(\text{H}^+) + \Lambda^\infty(\text{Cl}^-) + \Lambda^\infty(\text{CH}_3\text{COO}^-) + \Lambda^\infty(\text{Na}^+) - \Lambda^\infty(\text{Na}^+) - \Lambda^\infty(\text{Cl}^-)$$

$$= \Lambda^\infty(\text{H}^+) + \Lambda^\infty(\text{CH}_3\text{COO}^-)$$

$$= \Lambda^\infty(\text{CH}_3\text{COOH})$$

Hydrochloric acid, sodium acetate and sodium chloride are all strong electrolytes for which Λ^∞ can readily be determined experimentally.

DETERMINATION OF THE SOLUBILITY OF A SPARINGLY SOLUBLE SALT

If a sparingly soluble salt dissociates simply and completely, its solubility may be determined from conductance measurements.

The conductivity of a saturated solution of the salt is determined, subtracting that of the solvent. The result is the conductivity due to the salt itself:

$$\kappa_{\text{soln.}} - \kappa_{\text{solvent}} = \kappa_{\text{salt}}$$

Applying eqn. (2.7), we have

$$\Lambda_s = \frac{\kappa_{\text{salt}}}{c}$$

where Λ_s is the molar conductivity of the salt in the saturated solution and c the concentration of the solution. If the salt is only sparingly soluble, its saturated solution may be considered to be nearly infinitely dilute. We may thus make the approximation

$$\Lambda_s \approx \Lambda^\infty$$

whence

$$c = \frac{\kappa_{\text{salt}}}{\Lambda^\infty}$$

or, applying eqn. (2.16)

$$c = \frac{\kappa_{\text{salt}}}{\nu_+ \Lambda_+{}^\infty + \nu_- \Lambda_-{}^\infty} \tag{5.2}$$

Example—The conductivity of a saturated solution of silver chloride at 25°C is 3.41×10^{-6}, that of water at the same temperature is 1.60×10^{-6} Ω^{-1} cm^{-1}. Given that the limiting molar conductivities of the silver ion and the chloride ion are 61.92 and 76.34 Ω^{-1} cm^2 mol^{-1}, respectively, at 25°C, calculate the solubility of silver chloride at this temperature.

$$\kappa_{AgCl} = (3.41 \times 10^{-6}) - (1.60 \times 10^{-6})$$

$$= 1.81 \times 10^{-6} \ \Omega^{-1} \text{cm}^{-1}$$

Applying eqn. (5.2)

$$c = \frac{1.81 \times 10^{-6} \ \Omega^{-1} \text{cm}^{-1}}{(61.92 + 76.34) \ \Omega^{-1} \text{cm}^2 \text{mol}^{-1}}$$

$$= \frac{1.81 \times 10^{-6}}{138.26} \ \text{mol cm}^{-3}$$

$$= 1.31 \times 10^{-8} \ \text{mol cm}^{-3}$$

$$= 1.31 \times 10^{-5} \ \text{mol dm}^{-3}$$

This method has severe limitations. The saturated solution must be sufficiently dilute for its molar conductivity to be put equal to the limiting value without serious error. A further consideration is that the electrolyte must ionise simply and completely. If uncharged ion pairs occur in the solution, they will make no contribution to the conductance and erroneous results will be obtained. Similarly, if complex ions are formed, considerable error will be introduced.

DETERMINATION OF DISSOCIATION CONSTANTS

The determination of dissociation constants from conductance data can be quite complicated with regard to the treatment of the experimental data, especially from the stronger electrolytes. We will thus confine our considerations to the determination of the dissociation constants of weak electrolytes, and we shall start with a very simple treatment.

Consider a weak acid, HA, which dissociates according to

$$HA + H_2O \rightleftharpoons H_3O^+ + A^-$$

Its dissociation constant is given by eqn. (3.8)

$$K_a = \frac{[H^+][A^-]}{[HA]} \cdot \frac{y_{H^+} y_{A^-}}{y_{HA}}$$

If we consider only very dilute solutions, we may take the activity coefficients as equal to unity and write

$$K_a = \frac{[H^+][A^-]}{[HA]} \tag{5.3}$$

If the degree of dissociation of the acid is α at concentration c eqn. (5.3) may be written in the form

$$K_a = \frac{\alpha^2 c}{(1 - \alpha)} \tag{5.4}$$

Taking the degree of dissociation as equal to the conductance ratio

$$\alpha = \Lambda / \Lambda^\infty$$

and substituting into eqn. (5.4)

$$K_a = \frac{(\Lambda/\Lambda^\infty)^2 c}{(1 - \Lambda/\Lambda^\infty)}$$

$$= \frac{c\Lambda^2}{\Lambda^{\infty 2} (1 - \Lambda/\Lambda^\infty)}$$

Rearranging

$$c\Lambda = K_a \frac{\Lambda^{\infty 2}}{\Lambda} \left(1 - \frac{\Lambda}{\Lambda^\infty}\right)$$

$$= K_a \left(\frac{\Lambda^{\infty 2}}{\Lambda} - \Lambda^\infty\right) \tag{5.5}$$

From eqn. (5.5) it can be seen that a plot of $c\Lambda$ against $1/\Lambda$ should be a straight line of slope $K_a \Lambda^{\infty 2}$ and intercept $- K_a \Lambda^\infty$. The dissociation constant may thus be obtained.

It must be remembered that eqn. (5.5) is only valid for very dilute solutions, as we have assumed that the activity coefficients are unity and that the degree of dissociation is equal to the conductance ratio.

It is very difficult to obtain reliable conductance data at low concentrations, and the method based on eqn. (5.5) will give only approximate results. Conductance measurements should preferably be carried out at concentrations higher than the limits of applicability of this equation, where activity coefficients deviate from unity and where it is no longer justifiable to take the degree of dissociation as equal to the conductance ratio. It may be said that the complexity of the treatment of the experimental data in all accurate determinations of dissociation constants arises from these considerations.

This point may be illustrated by examining a more accurate method for determining the dissociation constant of a weak electrolyte, although we shall restrict ourselves to one in which the treatment of the data is still relatively simple.

The exact expression for the dissociation constant of a weak acid, HA, is given by eqn. (3.8). The expression

$$\frac{[H^+][A^-]}{[HA]}$$

is equal to the dissociation constant only for very dilute solutions, and in the more general case now considered it is better to write

$$k_a = \frac{[H^+][A^-]}{[HA]} \qquad (5.6)$$

where k_a is the 'classical dissociation constant' which varies slightly with concentration. Substituting into eqn. (3.8), we have

$$K_a = k_a \cdot \frac{y_{H^+} y_{A^-}}{y_{HA}} \qquad (5.7)$$

If the solutions are still reasonably dilute (0.1 mol dm^{-3}), the activity coefficient of the undissociated HA molecules will be near to unity, and eqn. (5.7) becomes

$$K_a = k_a(y_{H^+} y_{A^-}) \qquad (5.8)$$

As the H^+ and the A^- ions are both univalent, their activity coefficients will be equal, and representing them as y, eqn. (5.8) reduces to

$$K_a = k_a y^2$$

Taking logarithms

$$\log K_a = \log k_a + 2 \log y \qquad (5.9)$$

The term $\log y$ is given by the Debye–Hückel equation (3.2) which, in this case, takes the form

$$\log y = - A \sqrt{I} \qquad (5.10)$$

where I is the ionic strength of the solution, given by

$$I = \tfrac{1}{2}(c_{H^+} + c_{A^-})$$

c_{H^+} and c_{A^-} being the concentrations of the H^+ and A^- ions, respectively. If the concentration of the acid is c and the degree of dissociation α, the concentration of each ion must be αc

$$c_{H^+} = c_{A^-} = \alpha c$$

and hence

$$I = \tfrac{1}{2}(\alpha c + \alpha c) = \alpha c$$

Substituting in eqn. (5.10), we have

$$\log y = - A \sqrt{\alpha c}$$

and substituting again into eqn. (5.9)

$$\log K_a = \log k_a - 2A \sqrt{\alpha c}$$

or

$$\log k_a = 2A \sqrt{\alpha c} - \log K_a \qquad (5.11)$$

A plot of $\log k_a$ against $\sqrt{\alpha c}$ thus gives a straight line of intercept $\log K_a$. The value of $\log k_a$ is calculated from the relationship

$$k_a = \frac{\alpha^2 c}{(1 - \alpha)}$$

The treatment so far has accounted for the fact that the ionic activity coefficients will not be unity but, at the concentrations

143

employed, α cannot be calculated from the conductance ratio Λ/Λ^∞.

In Chapter 2 it was pointed out that the conductance ratio gives the degree of dissociation only on the assumption that the speeds of the ions do not change with concentration. This will only be true for very dilute solutions of weak electrolytes. If the solutions are other than very dilute, we must take into account the variation of ionic speed with concentration. The degree of dissociation is more accurately given by

$$\alpha = \frac{\Lambda}{\Lambda'}$$

where Λ' is the conductance which the electrolyte would have if it were fully dissociated at concentration c. Λ' may be calculated from an equation of the type of (5.1), namely

$$\Lambda'(HA) = \Lambda(HCl) + \Lambda(NaA) - \Lambda(NaCl)$$

where the Λ values for the three strong electrolytes all refer to the same concentration αc.

DETERMINATION OF HYDROLYSIS CONSTANTS

Consider the hydrolysis of aniline hydrochloride in aqueous solution. This is the salt of a weak base and a strong acid, and the anilinium ion will interact with the solvent

$$C_6H_5NH_3^+ + H_2O \rightleftharpoons C_6H_5NH_2 + H_3O^+$$

Assuming that the solution is sufficiently dilute for the activity coefficients is to be taken as unity, the hydrolysis constant will be given by

$$K_h = \frac{[C_6H_5NH_2]\,[H^+]}{[C_6H_5NH_3^+]}$$

If the concentration of aniline hydrochloride is c and the degree of hydrolysis α, this expression may be written as

$$K_h = \frac{\alpha^2 c}{(1 - \alpha)} \tag{5.12}$$

APPLICATIONS OF CONDUCTANCE MEASUREMENTS

The value of α may be determined from conductance measurements, and thus the hydrolysis constant can be calculated.

Suppose the molar conductivity of the solution is Λ. This conductivity will be due partly to the unhydrolysed anilinium ions and the chloride ions associated with them, and partly to the hydrogen ions produced by hydrolysis and the chloride ions which may now be considered to be associated with the hydrogen ions. The solution may be considered to consist of unhydrolysed aniline hydrochloride of concentration $c\,(1 - \alpha)$, free aniline of concentration αc and free hydrochloric acid of concentration αc.

If Λ' is the molar conductivity of unhydrolysed aniline hydrochloride at concentration c and Λ'' that of hydrochloric acid at concentration c, then

$$\Lambda = (1 - \alpha)\,\Lambda' + \alpha\Lambda'' \tag{5.13}$$

the contribution of the free aniline to the conductance of the solution being considered negligible. Equation (5.13) may be rearranged

$$\Lambda = \Lambda' + \alpha(\Lambda'' - \Lambda')$$

whence

$$\alpha = \frac{\Lambda - \Lambda'}{\Lambda'' - \Lambda'} \tag{5.14}$$

Λ is calculated from conductance measurements on solutions of aniline hydrochloride, and Λ' is obtained from the conductance of solutions in which hydrolysis has been almost completely suppressed by the addition of excess aniline. Knowing values of Λ'' from conductance measurements on hydrochloric acid solutions, it is possible to calculate α from eqn. (5.14) and, hence, K_h from eqn. (5.12). It will be remembered that K_h could equally well be regarded as the acid dissociation constant of the anilinium ion.

CONDUCTIMETRIC TITRATIONS

During the course of a titration, the concentrations of various ions in solution are changing. If these changes produce a net change in the conductance of the solution, the course of the

145

titration may be followed from conductance measurements. This technique finds its greatest application in the cases of acid–base titrations and precipitation titrations.

In order to understand the principles underlying conductimetric titrations, it is necessary to have some idea of the relative magnitudes of the conductances of ions. This information may be obtained from *Table 6* (p. 27) which shows that the limiting molar conductivities of most ions lie in the range 40–70 Ω^{-1} cm^2 mol^{-1}. The notable exceptions are hydrogen ions and hydroxide ions, for which the appropriate figures are about 350 and 198 Ω^{-1} cm^2 mol^{-1}, respectively. Although observations are not made at infinite dilution in conductimetric titrations, we may say that, with the exception of hydrogen ions and hydroxide ions, most other ions have similar conductances. Bearing these facts in mind, we may now examine particular examples of conductimetric titrations.

(*a*) *Strong acid–strong base*—Consider, for example, the titration of hydrochloric acid with sodium hydroxide. Originally the solution will contain hydrogen ions and chloride ions. When sodium ions and hydroxide ions are added in the form of the sodium hydroxide solution, the hydroxide ions will combine with the hydrogen ions to form water. At the end-point the solution will consist of sodium ions and chloride ions. The ions arising from the very slight dissociation of the water are ignored, as their contribution to the total conductance of the solution will be negligible compared with that due to the other ions present.

At the start of the titration, the solution contains hydrogen ions and chloride ions, and at the end-point, sodium ions and chloride ions. It is obvious that, as the titration proceeds towards the end-point, the highly conducting hydrogen ions are being replaced by sodium ions of much lower conductance. As a result of this the conductance of the solution will decrease.

After the end-point has been reached, the addition of further amounts of sodium hydroxide to the solution results only in the addition of extra ions, namely, sodium ions and hydroxide ions. The appearance of these extra ions will cause the conductance of the solution to increase.

If the conductance of the solution were to be plotted against the amount of titrant added, it would decrease to a minimum at the end-point, after which it would increase. The actual values of the conductivity of the solution need not be calculated, as only relative values are required, and the readings of the conductance bridge may be used directly without having to correct them with the cell constant. The values of conductances which may be used to plot a graph will thus be proportional to the conductivities of the solution.

If it were possible to add the alkali to the acid without changing the volume of the solution, the conductance would decrease linearly with the amount of alkali added and then increase in a very nearly linear manner after the end-point. The graph would then consist of two intersecting straight lines. In practice, the volume of the solution does increase, and this means that the same number of ions are accommodated in a larger volume, which will cause the conductance to decrease more rapidly still, thus causing a deviation from a linear relationship. This deviation may be largely compensated by multiplying the conductance reading by the total volume of the solution. Thus, if the conductance, K, multiplied by the total volume of the solution, V, is plotted against the volume of alkali added, v, the type of graph illustrated in *Figure 20* will be obtained.

(*b*) *Weak acid–strong base*—The data which would be obtained for the titration of a weak acid by a strong base are also shown in *Figure 20*. In the case of a weak acid, the original solution will not contain as many ions as a strong acid solution and the initial conductance will be much lower. Denoting the acid as HA, its ionisation may be represented as

$$HA \rightleftharpoons H^+ + A^-$$

The solution will contain H^+ ions, A^- ions and un-ionised HA molecules. Suppose that the solution is to be titrated with sodium hydroxide and consider the situation when only a small amount of alkali has been added. Hydrogen ions in the solution will have been replaced by sodium ions. Their removal will encourage more acid molecules to dissociate to restore the above equilibrium, but the dissociation will be suppressed by the presence of the A^- ions which still exist as free ions and may be considered as the anions of the salt NaA which will be a strong electrolyte and fully dissociated. Because of this suppression

147

of further dissociation of the acid, the net effect is that hydrogen ions have been replaced by sodium ions and the conductance of the solution suffers an initial decrease. As the titration continues, the process may be regarded as the replacement of un-ionised HA molecules by Na^+ ions and A^- ions, and the conductance of the solution begins to increase. Before the end-point, then, Na^+ and A^- ions are being added to the solution. After the end-point, the sodium hydroxide which is added is in excess and

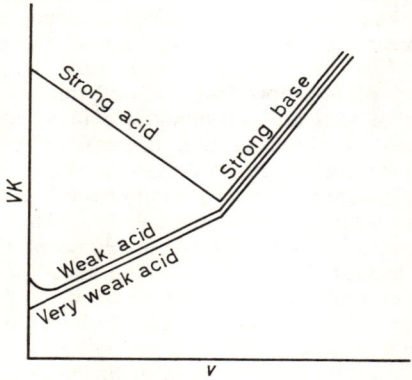

Figure 20. Conductimetric titration of acids with strong bases

V = *total volume of solution*
K = *conductance of solution*
v = *volume of base added*

remains in solution as Na^+ ions and OH^- ions. As the conductance of the OH^- ion is greater than that of the A^- ion, that of the solution continues to increase but at a greater rate after the end-point. This behaviour is illustrated in *Figure 20*.

As the titration is between a weak acid and a strong base, the salt which is formed will be subject to hydrolysis. Whilst excess acid or base is present, this will be suppressed and will only be apparent in the region of the end-point. The conductance plot will not show a sharp discontinuity at the end-point, but the transition between the two straight lines will appear as a smooth curve. The end-point may still be determined by producing the straight sections of the graph to the point of intersection.

(*c*) *Very weak acid–strong base*—When the acid to be titrated is very weak, the number of ions initially present may be considered to be negligible. When the alkali is added, the salt which is formed will be fully ionised and the titration may be regarded as the addition of ions to the solution. In this case, the conductance of the solution increases from the start of the titration and continues to increase more rapidly after the end-point has been passed. Once again, hydrolysis will cause the junction

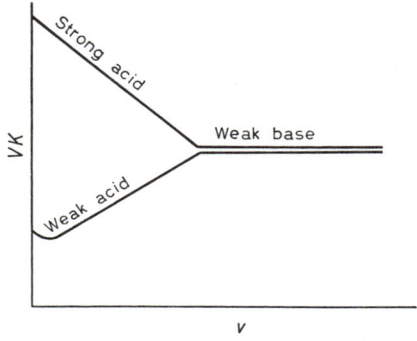

Figure 21. Conductimetric titration of acids with weak bases

V = *total volume of solution*
K = *conductance of solution*
v = *volume of base added*

between the two straight portions of the graph to be in the form of a smooth curve.

(*d*) *Acids–weak bases*—When acids are titrated with weak bases, the shapes of the graphs obtained before the end-points are exactly similar to those obtained when the titrant is a strong base. After the end-point has been reached, however, the excess base which is present in the solution will be largely un-ionised if the base is weak. The presence of excess base thus makes no significant contribution to the conductance of the solution which remains virtually constant at its end-point value. This behaviour is illustrated in *Figure 21*.

In the case of a strong acid–weak base titration, hydrolysis of the salt will once again cause some curvature at the point of intersection of the straight portions of the graph. In the case of a weak acid–weak base titration, hydrolysis of the salt will have a somewhat greater effect on the conductance plot, causing curvature of the graph over a larger region. In fact, to obtain a satisfactory end-point, the acid and base must be only moderately weak (i.e. K_a and K_b not less than 10^{-5}). If the acid and base are too weak, the graph may appear as a smooth curve over the whole range of observations.

(*e*) *Mixtures of acids*—Using a conductimetric technique, a mixture of a weak acid and a strong acid may be titrated with a base. When alkali is added to the acid solution, the first reaction which occurs is the replacement of the hydrogen ions which the strong acid provides by the cations of the alkali. The conductance of the solution thus falls in a manner corresponding to the neutralisation of a strong acid. Once the strong acid is completely neutralised, further amounts of alkali go to neutralising the weak acid. This process is essentially the replacement of un-ionised acid molecules by the ions of the salt formed, and the conductance of the solution increases. After the weak acid has been neutralised, the conductance of the solution will increase at a greater rate if a strong base has been used as the titrant. A plot of conductance against the amount of base added will thus give a graph with two breaks in it. The first corresponds to the titre for the strong acid, and the interval between the two to the titre for the weak acid. The situation is shown in *Figure 22*.

(*f*) *Precipitation titrations*—Consider the titration of a solution of potassium chloride with silver nitrate solution. The reaction occurring is the precipitation of silver chloride

$$KCl + AgNO_3 \rightarrow AgCl \downarrow + KNO_3$$

Initially, the solution consists of potassium ions and chloride ions. At the end-point the solution may be considered to consist of potassium ions and nitrate ions if the slight solubility of silver chloride is neglected. Up to the end-point, the net effect is the replacement of chloride ions by nitrate ions. *Table 6* shows that the conductances of these two ions are very similar, so that no appreciable change in the conductance of the solution will occur up to the end-point. After the end-point has been passed, the excess silver nitrate added merely increases the number of ions in solution, and the conductance will thus increase. The point at which this change occurs indicates the end-point of the titration.

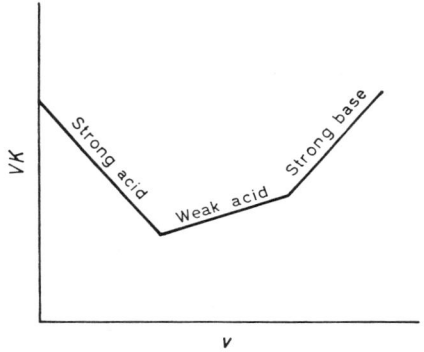

Figure 22. Conductimetric titration of acid mixtures

V = *total volume of solution*
K = *conductance of solution*
v = *volume of base added*

Consider now the titration of a solution of magnesium sulphate with barium hydroxide solution. During its course both magnesium hydroxide and barium sulphate will be precipitated

$$MgSO_4 + Ba(OH)_2 \rightarrow Mg(OH)_2 \downarrow + BaSO_4 \downarrow$$

The original solution contains magnesium ions and sulphate ions, and as the titration proceeds, both these ions are removed as precipitates. The conductance of the solution will thus decrease to very nearly zero at the end-point, when the only ions in solution are those due to the very slight solubilities of

151

magnesium hydroxide and barium sulphate. After the end-point has been passed, barium ions and hydroxide ions are being fed into the solution by the excess barium hydroxide, and the conductance of the solution increases again. The end-point of the titration is readily detected, as it corresponds to the minimum conductance.

The courses of the above two precipitation titrations are illustrated in *Figure 23*.

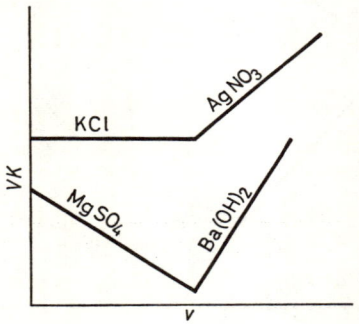

Figure 23. Conductimetric titration of precipitation reactions

V = total volume of solution
K = conductance of solution
v = volume of base added

6

APPLICATIONS OF E.M.F. MEASUREMENTS

DETERMINATION OF SOLUBILITY PRODUCTS

CONSIDER a cell in which a silver–silver chloride electrode is combined with a calomel reference electrode

$$Hg/Hg_2Cl_2(s), KCl//KCl, AgCl(s)/Ag$$
calomel reference electrode

By measuring the e.m.f. of this cell, the potential of the silver electrode may be obtained, provided that the potential of the calomel electrode is known. From eqn. (4.10)

$$E_{cell} = E_{Ag} - E_{cal}$$

or

$$E_{Ag} = E_{cell} + E_{cal}$$

The potential of the silver electrode is given by eqn. (4.12) as

$$E_{Ag} = E_{Ag}^° + \frac{RT}{F} \ln K_{AgCl} - \frac{RT}{F} \ln a_{Cl^-}$$

where a_{Cl^-} is the activity of the chloride ions in the silver–silver chloride electrode, and K_{AgCl} the solubility product of silver chloride. Rearranging the above equation

$$\ln K_{AgCl} = (E_{Ag} - E_{Ag}^°) \frac{F}{RT} + \ln a_{Cl^-}$$

or

$$\log K_{AgCl} = (E_{Ag} - E_{Ag}^°) \frac{F}{2·303RT} + \log a_{Cl^-} \qquad (6.1)$$

153

In order to calculate K_{AgCl} from eqn. (6.1), the standard electrode potential of silver would have to be known, and the activity of the chloride ions in the silver–silver chloride electrode would have to be assumed equal to the mean ionic activity of the potassium chloride. In this particular case, a salt bridge would be unnecessary, as the liquid junction is between two solutions of potassium chloride and its potential is thus negligible.

This method finds wide application despite the approximation that has to be made, as a suitable electrode system is fairly easy to devise for most cases.

An alternative cell may be set up in which the necessity of knowing the standard potential of the metal electrode is avoided. For example

$$Ag/AgCl(s), KCl//AgNO_3/Ag$$

If the potentials of the left- and right-hand electrodes are E_1 and E_2, respectively, the e.m.f. of the cell is given by

$$E_{cell} = E_2 - E_1$$

The potentials E_2 and E_1 are given by eqn. (4.5) and (4.12) as

$$E_2 = E_{Ag}^\circ + \frac{RT}{F} \ln (a_+)_2$$

$$E_1 = E_{Ag}^\circ + \frac{RT}{F} \ln K_{AgCl} - \frac{RT}{F} \ln (a_-)_1$$

where $(a_+)_2$ is the activity of the silver ions in the right-hand electrode, $(a_-)_1$ that of the chloride ions in the left-hand one. Hence

$$E_{cell} = \left(E_{Ag}^\circ + \frac{RT}{F} \ln (a_+)_2 \right)$$

$$- \left(E_{Ag}^\circ + \frac{RT}{F} \ln K_{AgCl} - \frac{RT}{F} \ln (a_-)_1 \right)$$

$$= \frac{RT}{F} \ln (a_+)_2 + \frac{RT}{F} \ln (a_-)_1 - \frac{RT}{F} \ln K_{AgCl}$$

Rearranging

$$\ln K_{AgCl} = \ln (a_+)_2 + \ln (a_-)_1 - \frac{FE_{cell}}{RT}$$

or

$$\log K_{AgCl} = \log (a_+)_2 + \log (a_-)_1 - \frac{FE_{cell}}{2 \cdot 303 RT} \tag{6.2}$$

Once again, $(a_+)_2$ has to be put equal to the mean ionic activity of the silver nitrate, and $(a_-)_1$ is assumed equal to that of the potassium chloride.

A more accurate method of determining the solubility product of silver chloride would be to measure the e.m.f. of a suitable cell without a liquid junction. Such a cell could be

$$Ag/AgCl(s), HCl/Cl_2 ; Pt$$
$$1 \text{ atm}$$

$$E_{cell} = E_{Cl_2} - E_{Ag}$$

The potentials of the individual electrodes are given by eqn. (4.7) and (4.12) as

$$E_{Cl_2} = E_{Cl_2}^\circ - \frac{RT}{F} \ln a_{Cl^-}$$

and

$$E_{Ag} = E_{Ag}^\circ + \frac{RT}{F} \ln K_{AgCl} - \frac{RT}{F} \ln a_{Cl^-}$$

Thus

$$E_{cell} = (E_{Cl_2}^\circ - E_{Ag}^\circ) - \frac{RT}{F} \ln K_{AgCl}$$

Rearranging

$$\ln K_{AgCl} = (E_{Cl_2}^\circ - E_{Ag}^\circ - E_{cell}) \frac{F}{RT}$$

or

$$\log K_{AgCl} = (E_{Cl_2}^\circ - E_{Ag}^\circ - E_{cell}) \frac{F}{2 \cdot 303 RT} \tag{6.3}$$

It will be noticed that, owing to the absence of a liquid junction, no assumptions have to be made, and eqn. (6.3) may be used without further modification to calculate K_{AgCl}. Despite the fact that this last method is more accurate than the first two, it is not so widely used, as suitable cells are more difficult to devise.

DETERMINATION OF THERMODYNAMIC FUNCTIONS

The relationships between ΔG, ΔH, ΔS and the e.m.f. of cells have already been derived in Chapter 4 and are given by eqn. (4.22), (4.23) and (4.24) as

$$\Delta G = - zFE$$

$$\Delta H = - zFE + zFT \left(\frac{\partial E}{\partial T} \right)_p$$

$$\Delta S = zF \left(\frac{\partial E}{\partial T} \right)_p$$

A numerical example will be considered here in order to illustrate the use of these equations.

For the cell

$$\text{Pb/PbCl}_2\text{(s), KCl, AgCl(s)/Ag}$$

the e.m.f. and its temperature coefficient at 25°C are 0·4902 V and $-0·000186 \text{ V K}^{-1}$, respectively. Write down the cell reaction and calculate the values of ΔG, ΔH and ΔS for the reaction at 25°C.

At the anode, (Pb), an oxidation reaction must occur between Pb, PbCl$_2$(s) and Cl$^-$ ions

$$\text{Pb} + 2\text{Cl}^- \rightarrow \text{PbCl}_2\text{(s)} + 2e$$

At the cathode, (Ag), a reduction reaction must occur between Ag, AgCl(s) and Cl$^-$ ions

$$\text{AgCl(s)} + e \rightarrow \text{Ag} + \text{Cl}^-$$

or

$$2\text{AgCl(s)} + 2e \rightarrow 2\text{Ag} + 2\text{Cl}^-$$

The overall cell reaction is therefore

$$Pb + 2AgCl(s) \rightarrow PbCl_2(s) + 2Ag$$

For this amount of reaction, $z = 2$ and hence

$$\Delta G = -2 \times 96\,500 \times 0.4902 \quad J \, mol^{-1}$$

$$= -96.6 \, kJ \, mol^{-1}$$

$$\Delta H = (-2 \times 96\,500 \times 0.4902)$$
$$+ [2 \times 96\,500 \times 298 \times (-0.000\,186)] \, J \, mol^{-1}$$

$$= -2 \times 96\,500 \, [0.4902 + (298 \times 0.000\,186)] \, J \, mol^{-1}$$

$$= -2 \times 96\,500 \times 0.5456 \, J \, mol^{-1}$$

$$= -105.3 \, kJ \, mol^{-1}$$

$$\Delta S = 2 \times 96\,500 \times (-0.000\,186) \, J \, K^{-1} \, mol^{-1}$$

$$= -35.9 \, J \, K^{-1} \, mol^{-1}$$

MEASUREMENT OF pH

One of the most important applications of e.m.f. measurements is the determination of the pH of a solution. It will be remembered that pH was formally defined in Chapter 3 as

$$pH = - \log a_{H^+}$$

If an electrode which enters into equilibrium with hydrogen ions can be found, its potential should be a measure of the pH of the solution.

(a) *The hydrogen electrode*—The obvious choice of an electrode which responds to the activity of hydrogen ions in solution is the hydrogen electrode. The electrode equilibrium is

$$\tfrac{1}{2}H_2 \rightleftharpoons H^+ + e$$

and its potential is obtained by applying eqn. (4.1). As the standard potential of a hydrogen electrode is zero by convention, we have

$$E = \frac{RT}{F} \ln \frac{a_{H^+}}{a_{H_2}^{1/2}} \qquad (6.4)$$

Assuming that the activity of the gas is equal to its partial pressure, p, eqn. (6.4) becomes

$$E = \frac{RT}{F} \ln a_{H^+} - \frac{RT}{F} \ln p^{1/2} \qquad (6.5)$$

Equation (6.5) may be further simplified if the partial pressure of the gas is 1 atm and

$$E = \frac{RT}{F} \ln a_{H^+}$$

$$= \frac{2 \cdot 303 RT}{F} \log a_{H^+} \qquad (6.6)$$

The hydrogen electrode is the primary standard of pH measurement and can be used in aqueous solutions over a wide range of pH. Precautions must be taken to ensure the purity of the supply of hydrogen gas and it must be remembered that the electrode cannot be used in solutions containing reducible substances. The electrode is also subject to 'poisoning' by substances such as sulphur, arsenic, cyanides, etc. Provided these limitations are observed, the hydrogen electrode gives reproducible and reliable results.

(b) *The quinhydrone electrode*—Haber and Russ first established the fact that quinone and hydroquinone constitute a redox system in which the equilibrium is dependent upon hydrogen ion activity. The equilibrium may be represented as

$$C_6H_4O_2 + 2H^+ + 2e \rightleftharpoons C_6H_4(OH)_2$$

Denoting the activities of quinone and hydroquinone as a_Q and a_{HQ}, respectively and applying eqn. (4.1) to the system

$$E = E^\circ + \frac{RT}{2F} \ln \frac{a_Q \times a_{H^+}{}^2}{a_{HQ}}$$

$$= E^\circ + \frac{RT}{2F} \ln \frac{a_Q}{a_{HQ}} + \frac{RT}{F} \ln a_{H^+} \qquad (6.7)$$

Quinone and hydroquinone form a 1:1 molecular compound called quinhydrone, $C_6H_4O_2 . C_6H_4(OH)_2$, which is sparingly soluble in water. A saturated solution of quinhydrone is thus very dilute and the activity of the quinone may be taken as equal to that of the hydroquinone. Under these conditions, eqn. (6.7) reduces to

$$E = E^\circ + \frac{RT}{F} \ln a_{H^+}$$

$$= E^\circ + \frac{2 \cdot 303 RT}{F} \log a_{H^+} \qquad (6.8)$$

The potential of the electrode would thus depend directly upon the hydrogen ion activity.

The quinhydrone electrode is set up by immersing a platinum or gold electrode in the solution of interest and adding a small quantity of quinhydrone. Good results are obtained in the pH range 1–8. In alkaline solutions, hydroquinone is subject to atmospheric oxidation which causes the ratio a_Q/a_{HQ} to deviate from unity and leads to the breakdown of eqn. (6.8) above pH 8. Another consideration is the fact that hydroquinone is itself an acid, and although it is very weak, its dissociation may become significant in alkaline solutions. The dissociation will provide secondary equilibria to which the potential of the electrode will respond, and the hydrogen ions arising from the dissociation may displace the pH of the original solution.

A further limitation of the quinhydrone electrode is the chemical reactivity of quinone. It reacts with amino compounds, ammonia and ammonium ions, and the electrode cannot be used in solutions containing these substances.

The electrode is also unreliable in solutions of high ionic strength when ' salt errors ' arise. With increasing ionic strength, the activities of quinone and hydroquinone will vary, but unfortunately to different extents. The activity ratio, a_Q/a_{HQ}, will again deviate from unity and eqn. (6.8) does not then apply. Salt errors become significant only above ionic strengths of 0.1 mol dm^{-3}, and below this limit the electrode gives reliable results.

(c) *The glass electrode*—Probably the most widely used indicator electrode in the measurement of pH is the glass electrode. It is

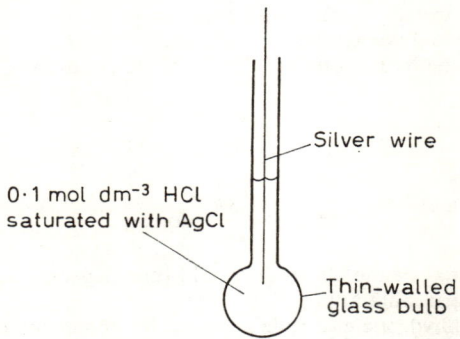

0.1 mol dm^{-3} HCl saturated with AgCl

Silver wire

Thin-walled glass bulb

Figure 24. Glass electrode

based on the observation of Haber and Klemensiewicz that the potential difference between the surface of a glass membrane and a solution varies linearly with the pH of the solution.

In its usual form the glass electrode consists of a tube terminating in a thin-walled bulb containing a solution of constant pH and an electrode of constant potential. Most frequently, the solution is 0.1 mol dm^{-3} hydrochloric acid and the electrode is a silver-silver chloride electrode (*Figure 24*).

In order to measure pH, the glass electrode is immersed in the experimental solution, so that the system may be represented as

Ag/AgCl(s), 0.1 mol dm^{-3} HCl/glass/experimental solution

The potential of the silver–silver chloride electrode will be constant, as it is in hydrochloric acid of constant activity. The potential between the inner surface of the glass and the hydrochloric acid solution will also be constant, as the pH of this solution does not change. The potential between the outer surface of the glass and the experimental solution will depend upon the pH of the solution, so that the overall potential of the system varies only with the pH of the experimental solution.

The most suitable glass for the purpose used to be Corning 015 glass which contains about 72 per cent SiO_2, 22 per cent Na_2O and 6 per cent CaO. This glass has now been largely superseded by lithia multicomponent glasses. Glass electrodes have a relatively low electrical resistance, but even so the resistance may be between 10 and 100 megohms ($M\Omega$). Due to this fact, special methods have to be employed for measuring the e.m.f. of any cell containing a glass electrode. It can easily be shown that the normal potentiometer method of measuring e.m.f. is inaccurate for cells of high internal resistance.

Referring to *Figure 17* (p. 100), it is seen that the e.m.f. of the experimental cell is balanced against that provided by the bridge wire between the point A and the sliding contact. The balance point is detected by the galvanometer which shows that no current flows round the circuit when the two e.m.f. are exactly balanced.

Suppose now that the internal resistance of the experimental cell is 10 $M\Omega$ and also that the sliding contact is not exactly at the balance point but is 1 mV out of balance. Normally, an out-of-balance condition is detected by a deflection on the galvanometer indicating a flow of current round the circuit. In the present example an excess voltage of $10^{-3}V$ is applied across a resistance of $10^7\Omega$. The current flowing through the galvanometer is given by Ohm's law

$$I = \frac{U}{R} = \frac{10^{-3}V}{10^7\Omega} = 10^{-10} \text{ A}$$

In order to indicate that the bridge is out of balance, the galvanometer must be capable of detecting a current of $10^{-10}A$. If the galvanometer can only just detect $10^{-10}A$, then the e.m.f. can only be measured to an accuracy of 1 mV. It can be understood, then, that the higher the resistance of the cell the more inaccurate does the measurement of its e.m.f. become.

The ordinary galvanometer used in conjunction with potentiometers can detect about 10^{-7}A, which is not good enough for systems involving glass electrodes. For the measurement of the e.m.f. of high-resistance cells, a valve amplifier circuit is usually employed to amplify the out-of-balance current before it is applied to the galvanometer. With such circuits, potentials may be measured to 0·5 mV which corresponds to 0·01 pH units.

Several theories have been advanced to account for the mechanism by which the potential of a glass surface depends upon the pH of the solution with which it is in contact. The theories are mostly complex and none of them alone account for all the aspects of behaviour of the glass electrode. It is probable that the pH response of the electrode is due to the entry of hydrogen ions into the lattice of the glass. One common feature of all the theories, however, is that they all yield the same result for the dependence of the potential upon pH; it reduces to a type of Nernst equation which may be written

$$E_G = k + \frac{RT}{F} \ln a_{H^+} \qquad (6.9)$$

where a_{H^+} is the activity of the hydrogen ions in the solution outside the electrode. The factor k is not a true constant and varies from day to day for a particular electrode. This is because it includes the *asymmetry potential* of the glass electrode which varies with time. If identical solutions and electrodes are placed inside and outside the glass electrode, the following cell will be formed

Ag/AgCl(s), 0·1 mol dm^{-3} HCl/glass/0·1 mol dm^{-3} HCl,
AgCl(s)/Ag

The e.m.f. of this cell should be zero, as it is perfectly symmetrical but in practice it is found to have an e.m.f. in the range 0–10 mV. This is due to differences in the pH response of the inner and outer surfaces of the glass electrode and is called the asymmetry potential of the electrode. It is thought to be due to different states of strain in the inner and outer surfaces of the glass. Thus the glass electrode has to be standardised with buffer solutions of known pH in order to determine the value of k in eqn. (6.9). It is essential to use at least two, and preferably more, to ensure that the value of k is constant over the range of pH in which measurements are to be made. The asymmetry potential varies with the treatment of the electrode and thus drifts slowly with time. It is therefore advisable to standardise a glass electrode at least once an hour during a long series of pH measurements.

A further reason for standardisation is that a plot of E_G against ln a_{H^+} should have a slope of RT/F but in practice deviations from this value may occur.

The glass electrode has many advantages. It can be used in solutions of oxidising and reducing agents, it does not contaminate the experimental solution and it is not affected by the usual electrode poisons. Certain proteins, however, will give rise to erroneous results, probably due to coagulation on the glass surface.

The ordinary glass electrode gives good results in the pH range 1–9 at temperatures less than 35°C. Above pH 9, the potential of the electrode is affected by other cations in solution in addition to hydrogen ions. These errors in pH response are probably due to penetration of the glass lattice by the other cations. The effect is particularly marked in the case of high concentrations of sodium ions. There are electrodes made of special glasses for measuring pH values greater than 9. In very acid solutions of pH < 1, the glass electrode is also subject to errors. It seems that the acid error is not due to specific cations or anions, although it may sometimes be due to adsorption of anions on the glass surface.

(d) *The antimony electrode*—Another electrode the potential of which responds to the pH of a solution is the antimony electrode. It is formed by casting a stick of antimony in air, so that some oxide is included. The equilibrium in solution may be written

$$Sb_2O_3 + 6H^+ + 6e \rightleftharpoons 2Sb + 3H_2O$$

The potential of the electrode is given by an equation of the type

$$E = A + B \log a_{H^+}$$

163

where A and B are constants depending upon the method of preparation of the electrode and the circumstances in which it is used. It is never used in laboratory measurements, as it is unreliable and must be calibrated for each different type of solution in which it is to be used. The main application of antimony electrodes is in industrial pH control where the pH of a particular type of solution has to be checked continually. Under these circumstances, it can be calibrated for a particular application and forms a robust electrode suitable for use in chemical plant.

MODERN DEFINITION OF pH

Although the pH of a solution is formally defined as

$$pH = - \log a_{H^+}$$

and although the potentials of the indicator electrodes, hydrogen, quinhydrone and glass, depend upon the activity of hydrogen ions in solution, the impossibility of measuring the potential of a single electrode precludes the determination of a pH value for a solution as defined above. The modern definition of pH stems from its method of measurement in which a suitable indicator electrode is combined with a reference electrode and the e.m.f. of the cell so formed is determined. The choice of indicator electrode is immaterial, as they all respond to pH in the same way, but for simplicity the hydrogen electrode will be considered.

Suppose that the pH of an unknown solution, X, is to be measured by immersing a hydrogen electrode in the solution and combining it with a calomel reference electrode. The complete cell may be represented by

$$Pt; H_2/\text{solution } X \mid KCl, Hg_2Cl_2(s)/Hg$$

where the dotted line represents the liquid junction between solution X and the potassium chloride solution of the calomel electrode. The e.m.f. of the complete cell, E^x, will be equal to the difference between the potentials of the electrodes plus any liquid junction potential, E_l^x

$$E^X = E_{Hg_2Cl_2} - E_{H_2} + E_l^X \qquad (6.10)$$

The potential of the calomel electrode is given by

$$E_{Hg_2Cl_2} = E_{Hg_2Cl_2}^{\circ} - \frac{RT}{F} \ln (a_{Cl^-})_R$$

where $(a_{Cl^-})_R$ is the activity of the chloride ions in the calomel reference electrode. Assuming that the partial pressure of hydrogen gas is 1 atm, the potential of the hydrogen electrode is given by

$$E_{H_2} = \frac{RT}{F} \ln (a_{H^+})_X$$

where $(a_{H^+})_X$ is the activity of the hydrogen ions in solution X. Substituting in eqn. (6.10)

$$E^X = E_{Hg_2Cl_2}^{\circ} - \frac{RT}{F} \ln (a_{Cl^-})_R - \frac{RT}{F} \ln (a_{H^+})_X + E_l^X$$

$$(6.11)$$

As it is not possible to measure the liquid junction potential or the activity of the chloride ions, it is impossible to obtain the value of $(a_{H^+})_X$ from eqn. (6.11).

Suppose now that the unknown solution X is removed from the cell and replaced with a standard solution, S, of known pH. The cell is now

$$\text{Pt; } H_2/\text{solution } S \;\vert\; \text{KCl, Hg}_2\text{Cl}_2(s)/\text{Hg}$$

The e.m.f. of this cell E^S will be given by an expression similar to eqn. (6.11)

$$E^S = E_{Hg_2Cl_2}^{\circ} - \frac{RT}{F} \ln (a_{Cl^-})_R - \frac{RT}{F} \ln (a_{H^+})_S + E_l^S$$

$$(6.12)$$

where $(a_{H^+})_S$ is the activity of the hydrogen ions in solution S and E_l^S the liquid junction potential. Subtracting eqn. (6.12) from (6.11)

$$E^X - E^S = \frac{RT}{F} \ln (a_{H^+})_S - \frac{RT}{F} \ln (a_{H^+})_X + (E_l^X - E_l^S)$$

$$(6.13)$$

Ignoring any difference between the liquid junction potentials $E_l{}^X$ and $E_l{}^S$ and assuming that the temperature of each cell is the same, eqn. (6.13) reduces to

$$E^X - E^S = \frac{RT}{F} \left(\ln (a_{H^+})_S - \ln (a_{H^+})_X \right)$$

$$= \frac{2 \cdot 303 RT}{F} \left(\log (a_{H^+})_S - \log (a_{H^+})_X \right)$$

$$= \frac{2 \cdot 303 RT}{F} \left(pH_X - pH_S \right)$$

Rearranging

$$pH_X = pH_S + \frac{F(E^X - E^S)}{2 \cdot 303 RT} \qquad (6.14)$$

Equation (6.14) is taken as the modern definition of pH. It is an *operational definition* in so far as any unknown solution can be placed in the type of cell described above and the e.m.f. of the cell can be measured. The procedure can be repeated with a standard solution which has had a value of pH_S assigned to it, and by application of eqn. (6.14) a number can be obtained for pH_X. This number may or may not have a *theoretical* significance. It can have some theoretical significance only if it is justifiable to ignore the difference between the liquid junction potentials in eqn. (6.13). When the two solutions, X and S, are very similar, the ionic strengths of both are less than $0 \cdot 1 \text{ mol dm}^{-3}$ and their pH is between 2 and 12, it is thought that the liquid junction potentials, $E_l{}^s$ and $E_l{}^x$, are nearly equal. Under these conditions, a pH number may have approximately the significance prescribed by the *formal* definition

$$pH = - \log a_{H^+}$$

Even so, the significance is doubtful, as the pH of the standard solution is assigned by convention.

ION SELECTIVE ELECTRODES

It was mentioned in the section on the glass electrode that at high pH the potential of the electrode was affected by other cations in addition to hydrogen ions, the effect being particulary marked in the case of high concentrations of sodium ions. This is a case of sodium ions interfering with a hydrogen responsive glass electrode. This led to the idea that if the sodium response could be increased and the hydrogen response diminished an electrode would be available which could give a measure of the concentration of sodium ions in solution. Glass electrodes have now been developed for the determination of potassium, ammonium and silver ions in addition to hydrogen and sodium ions. Such electrodes are called *ion selective electrodes*. These electrodes have been further developed and many are based on membranes other than glass to separate the internal reference solution from the experimental solution. The electrodes may be classified in four groups according to the type of membrane employed.

(1) *Glass electrodes.* These have been mentioned above and are cation selective only.

(2) *Solid state electrodes.* In these electrodes the membrane consists of a single crystal or a compacted disc of the active material. For example, a solid state electrode, selective to fluoride ions employs a membrane of lanthanum fluoride (LaF_3). One which is selective to sulphide ions has a membrane of silver sulphide. In addition to solid state electrodes for the determination of F^- and S^{2-} ions, there are also electrodes available for Cl^-, Br^-, I^-, Ag^+, Cu^{2+}, Pb^{2+}, Cd^{2+} and CN^- ions.

(3) *Heterogeneous membrane electrodes.* These electrodes are similar to the solid state electrodes but differ in having the active material dispersed in an inert matrix. In this class, electrodes are available for the measurement of Cl^-, Br^-, I^-, S^{2-} and Ag^+ ions.

(4) *Liquid ion exchange membrane electrodes.* In this type of electrode the internal reference solution and the experimental solution are separated by an organic liquid of low water solubility. Dissolved in the organic phase are large molecules in which the ion of interest is incorporated. The most important of these electrodes is the calcium electrode in which the ion exchanger is the calcium salt of didecylphosphoric acid dissolved in di-n-acetylphenylphosphonate. Other electrodes in this class are available for the determination of the ions Cl^-, ClO_4^-, NO_3^-, Cu^{2+}, Pb^{2+} and BF_4^-.

POTENTIOMETRIC TITRATIONS

Measurement of the potential of certain electrodes offers a convenient and accurate means for determining the end-points of titrations.

(a) *Acid–base titrations*—*Figure 12* (p. 82) shows how the pH of the solution varies during the course of acid–base titrations. It has been shown that the potential of a hydrogen electrode depends upon pH in the manner defined by eqn. (6.6) which may be written in the form

$$E = -\frac{2 \cdot 303 RT}{F} \, \text{pH} \qquad (6.15)$$

It can be seen that, as the pH of the solution increases when alkali is added to an acid, so the potential of a hydrogen electrode decreases. As the pH of the solution changes rapidly at the end-point of the titration, so does the potential of the hydrogen electrode. That of a single electrode cannot be measured, so the hydrogen electrode has to be combined with a suitable reference electrode, and the e.m.f. of the complete cell is measured. The potential of the reference electrode remains constant, so that the variation in the e.m.f. of the cell is due solely to that of the hydrogen electrode potential. If a saturated calomel electrode is used as a reference electrode, the observed e.m.f., E, will be given by

$$E = E_{Hg_2Cl_2} - E_{H_2}$$

As the potential of the hydrogen electrode decreases during the titration, the observed e.m.f. will increase. *Figure 25* shows a plot of the e.m.f. of a hydrogen–calomel combination for the titration of a strong acid with a strong base. The end-point of the titration is indicated by the vertical step in the graph.

As the quinhydrone and glass electrodes also respond to pH, they may be used as indicator electrodes equally as well as a hydrogen electrode. In practice, the most commonly used combination is a glass electrode with a saturated calomel reference electrode, both of which are immersed directly in the titration mixture.

(b) *Precipitation titrations*—Consider the titration of potassium chloride with silver nitrate. As soon as the first drop of silver nitrate is added, silver chloride will be precipitated and the solution will be saturated with respect to it. The activity of silver ions will be very small and may be indicated by the potential of a silver electrode immersed in the solution. As more silver nitrate is added, more chloride ions are precipitated as silver chloride. The solution remains saturated with silver chloride, but the activity of the silver ions will increase very slightly as

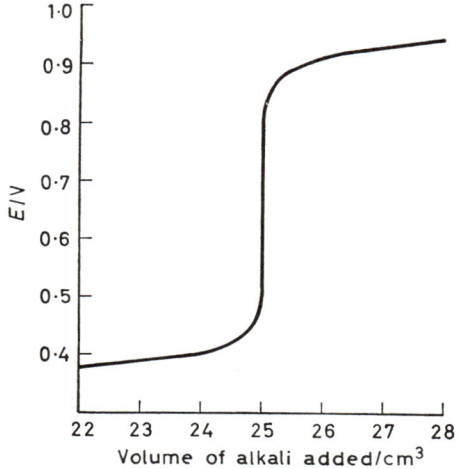

Figure 25. Potentiometric acid–base titration

chloride ions are removed, because the solubility product principle states

$$a_{Ag^+} \times a_{Cl^-} = K_{AgCl}$$

When the end-point is reached, the activity of the silver ions in solution increases rapidly, owing to the presence of excess silver nitrate, and the potential of the silver electrode shows a correspondingly sharp increase. This behaviour is illustrated in *Figure 26*.

169

Once again, the silver electrode has to be coupled with a reference electrode and the e.m.f. of the complete cell is measured. The potential of the reference electrode remains constant and the e.m.f. of the cell changes in exactly the same way as the potential of the silver electrode. The end-point is indicated by the sharp rise in the curve of e.m.f. plotted against the volume of titrant added.

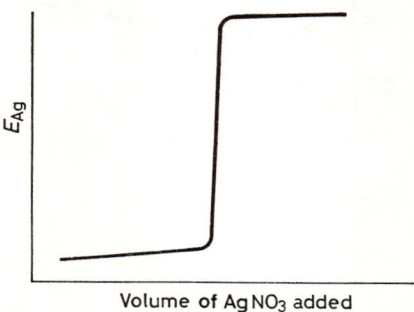

Figure 26. Potentiometric halide titration

Mixtures of halides and other anions which form insoluble silver compounds may also be titrated potentiometrically, with certain reservations. Consider a solution containing a mixture of chloride ions and iodide ions. When silver nitrate is added, silver iodide will be precipitated first, as it is the more insoluble of the two silver halides. During this process, the silver electrode potential will rise slightly as the iodide ions are removed from solution. When all the iodide has been precipitated, silver chloride will next be formed, and the potential of the silver electrode rises sharply to a new level which is dependent upon the solubility of silver chloride. During the precipitation of the chloride the potential will rise slightly and will then increase sharply at the end-point.

Figure 27 shows the type of curve which would be obtained. The first vertical step in the graph indicates the iodide titre and the interval between the first and second vertical steps the chloride titre. It is obvious that the height of the first step is proportional to the difference between the solubilities of silver iodide and silver chloride. It is therefore possible to titrate mixtures of anions only if there exist sufficient differences in the solubilities of their silver compounds to give observable steps in the graphs.

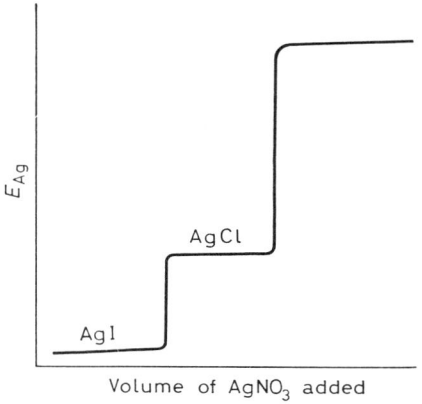

Figure 27. Potentiometric titration of mixed halides

If a saturated calomel electrode is to be used as a reference electrode, it cannot be directly immersed in the titration mixture because chloride ions from the calomel electrode would diffuse into the solution. The calomel electrode would be immersed in a beaker containing saturated potassium chloride or saturated ammonium nitrate, connected to the titration mixture by an ammonium nitrate salt bridge.

(*c*) *Redox titrations*—The variation of potential which would be shown by an inert electrode such as platinum during the course of a redox titration is indicated by *Figure 19* (p. 135). It can readily be appreciated from this diagram that a measurement of

171

the potential of an inert electrode with respect to a suitable reference electrode could be used to indicate the end-point of a redox titration.

(*d*) *Differential titrations*—An examination of E–v curves, as exemplified by *Figures 19*, *25* and *26*, shows that the end-point of a titration occurs at the vertical step. This is the point where the slope of the graph is a maximum. If a further plot of the slope of the E–v curve against the volume of titrant added were made,

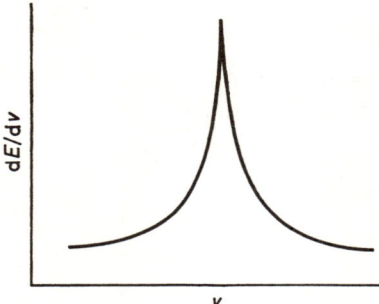

Figure 28. Differential plot

the type of graph illustrated in *Figure 28* would be obtained. The end-point of the titration now corresponds to the cusp in the differential plot.

A differential plot may be obtained directly from experimental observations by use of the simple apparatus illustrated in *Figure 29*. Two identical electrodes dip into the solution, but one is surrounded by a glass sheath from which the solution may be expelled by squeezing a rubber bulb. At the beginning of the titration, the solution is uniform throughout and both electrodes are surrounded by solution of the same composition. The potentials of the electrodes will thus be the same and the e.m.f. of the cell will be zero. Consider the situation when a small volume, Δv, of titrant has been added. The solution is stirred and the titrant reacts with the bulk of the solution but will not

172

affect the solution in the glass sheath. The electrodes are now surrounded by solutions of different compositions and there will exist a difference of potential ΔE between them. The ratio of $\Delta E/\Delta v$ will be approximately equal to dE/dv, especially if Δv is small. After the e.m.f. of the cell has been determined, the solution inside the glass sheath is expelled and allowed to mix

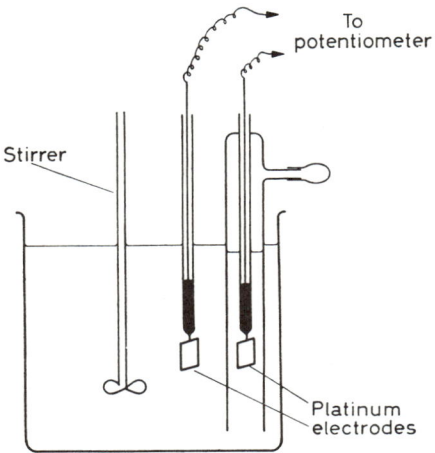

Figure 29. Differential potentiometric titration apparatus

freely with the bulk of the solution before the rubber bulb is released and some solution is drawn into the sheath again. The electrodes are once again surrounded by identical solutions and the process is repeated. If required, a differential graph of the type illustrated in *Figure 28* may be plotted, but this is not absolutely necessary.

At the beginning of the titration the values of ΔE will be small, as the difference in the potentials of the two electrodes is given by two points on *Figure 26*, say, which are separated by a volume Δv. At the end-point of the titration, however, the potential of the inner electrode may be at the bottom of the vertical step, whilst that of the outer electrode may be at its top, resulting in a large value of ΔE. The titration is thus continued until a maximum value of ΔE is obtained. The volumes of titrant added may be fairly large at the beginning of the titration, but as the end-point is approached it is desirable to make Δv as small as possible.

DETERMINATION OF TRANSPORT NUMBERS

The e.m.f., E_1, of the cell

$$\text{Zn/ZnSO}_4 \mid \text{ZnSO}_4/\text{Zn}$$
$$(a_\pm)_1 \quad (a_\pm)_2$$

is given by eqn. (4.34) as

$$E_1 = t_- \frac{RT}{F} \ln \frac{(a_\pm)_2}{(a_\pm)_1} \tag{6.16}$$

where t_- is the transport number of the sulphate ion. As transport numbers vary with concentration, the value of t_- in eqn. (6.16) will be the average of the transport numbers of the sulphate ion in the two solutions in the cell.

The e.m.f., E_2, of the cell

$$\text{Zn/ZnSO}_4, \text{Hg}_2\text{SO}_4(\text{s})/\text{Hg}/\text{Hg}_2\text{SO}_4(\text{s}), \text{ZnSO}_4/\text{Zn}$$
$$(a_\pm)_1 \qquad\qquad\qquad (a_\pm)_2$$

is given by eqn. (4.46) as

$$E_2 = \frac{RT}{F} \ln \frac{(a_\pm)_2}{(a_\pm)_1} \tag{6.17}$$

Dividing eqn. (6.16) by (6.17)

$$\frac{E_1}{E_2} = t_- \tag{6.18}$$

The transport number of the sulphate ion may thus be determined from measurements of the e.m.f. of the two cells given above. As the transport number obtained is an average value, the best results are achieved when the mean ionic activities of the zinc sulphate solutions differ by only a small amount.

MEAN IONIC ACTIVITY COEFFICIENTS

When dealing with ionic equilibria in Chapter 3 it was shown that the activity of an ion could be expressed as the product of an activity coefficient and the concentration of the ion

$$a = yc$$

where c is the actual ionic concentration, due allowance being made for incomplete dissociation. The ionic concentration in this equation is expressed as the *amount of substance divided by the volume of the solution*.

When dealing with electrochemical cells, it is the practice to deal with the *molality* of a solution which *is the amount of substance divided by the mass of the solvent*. The molality scale has the advantage of being independent of temperature and pressure, as it is based on weights. In electrochemical cell problems, the activity of an ion is usually expressed as the product of an activity coefficient and a molality. Moreover, the molality of the ion is calculated on the assumption that the electrolyte is completely dissociated, regardless of whether or not this is the case. For this reason it may be called the stoichiometric molality of the ion. The activity of an ion of type i may thus be written as

$$a_i = \gamma_i m_i \tag{6.19}$$

where m_i is the stoichiometric molality of the ion. The activity coefficient, γ_i, will not only account for ionic interaction but also for any incomplete dissociation of the electrolyte. For this reason, γ_i is sometimes called the *stoichiometric activity coefficient*.

The mean ionic activity of an electrolyte which dissociates into ν_+ cations and ν_- anions has already been defined by the relation

$$a_{\pm}^{\nu} = a_+^{\nu_+} . a_-^{\nu_-} \tag{6.20}$$

where $\nu = \nu_+ + \nu_-$

175

Applying eqn. (6.19) to the ions

$$a_+ = \gamma_+ m_+$$
$$a_- = \gamma_- m_-$$

Substituting in eqn. (6.20)

$$(a_\pm)^\nu = (\gamma_+{}^{\nu+} . \gamma_-{}^{\nu-})(m_+{}^{\nu+} . m_-{}^{\nu-}) \qquad (6.21)$$

The *mean ionic activity coefficient*, $\gamma_\pm$, is now defined as

$$\gamma_\pm{}^\nu = \gamma_+{}^{\nu+} . \gamma_-{}^{\nu-}$$

and the *mean ionic molality*, $m_\pm$, as

$$m_\pm{}^\nu = m_+{}^{\nu+} . m_-{}^{\nu-}$$

Equation (6.21) may thus be written

$$a_\pm = \gamma_\pm m_\pm \qquad (6.22)$$

DETERMINATION OF MEAN IONIC ACTIVITY COEFFICIENTS

Consider the cell

$$Pt; H_2/HCl, AgCl(s)/Ag$$

where the molality of the hydrochloric acid is m.

If the hydrogen gas is at a partial pressure of 1 atm, the potential of the hydrogen electrode, E_{H_2}, is given by

$$E_{H_2} = \frac{RT}{F} \ln a_+ \qquad (6.23)$$

where a_+ is the activity of the hydrogen ions in solution. The potential of the silver–silver chloride electrode, E_{AgCl}, is

$$E_{AgCl} = E_{AgCl}{}^\circ - \frac{RT}{F} \ln a_- \qquad (6.24)$$

where a_- is the activity of the chloride ions in solution.

The e.m.f. of the cell, E_{cell}, may be written as

$$E_{cell} = E_{AgCl} - E_{H_2}$$

Substituting from eqn. (6.23) and (6.24)

$$E_{cell} = E_{AgCl}^{\bullet} - \frac{RT}{F} \ln a_- - \frac{RT}{F} \ln a_+$$

$$= E_{AgCl}^{\bullet} - \frac{RT}{F} \ln a_+ a_- \qquad (6.25)$$

From eqn. (6.20)

$$a_+ a_- = a_{\pm}^2$$

where $a_{\pm}$ is the mean ionic activity of the hydrochloric acid. Equation (6.25) thus becomes

$$E_{cell} = E_{AgCl}^{\bullet} - \frac{2RT}{F} \ln a_{\pm}$$

or, applying eqn. (6.22)

$$E_{cell} = E_{AgCl}^{\bullet} - \frac{2RT}{F} \ln m_{\pm} \gamma_{\pm} \qquad (6.26)$$

For hydrochloric acid

$$m_{\pm}^2 = m_+ m_-$$

Moreover, the stoichiometric molalities of the hydrogen and chloride ions will be equal to that of the hydrochloric acid because it is a uni-univalent electrolyte. Hence

$$m_+ = m_- = m$$

and

$$m_{\pm}^2 = m^2$$

Equation (6.26) may now be written

$$E_{cell} = E_{AgCl}^{\bullet} - \frac{2RT}{F} \ln m\gamma_{\pm} \qquad (6.27)$$

If the standard potential of the silver–silver chloride electrode is known, the mean ionic activity coefficients may be determined from measurements of the e.m.f. of the cell at various molalities of hydrochloric acid.

DETERMINATION OF STANDARD ELECTRODE POTENTIALS

The accurate determination of standard electrode potentials is usually carried out with a cell having no liquid junction, as exemplified by that considered in the preceding section. The e.m.f. of the cell is given by eqn. (6.27) which may be written as

$$E_{cell} = E_{AgCl}^\circ - \frac{2RT}{F}\ln m - \frac{2RT}{F}\ln \gamma_\pm$$

or, converting to common logarithms

$$E_{cell} = E_{AgCl}^\circ - \frac{4\cdot606RT}{F}\log m - \frac{4\cdot606RT}{F}\log \gamma_\pm$$

$$(6.28)$$

The mean ionic activity coefficient now has to be expressed in terms of the concentration of the solution by means of a Debye–Hückel type of equation. The simple Debye–Hückel equation, since it is applicable only in very dilute solution, cannot be used, and as impurities can disturb the potentials of the electrodes in dilute solutions, meaningful measurements are not possible in this region. For accurate e.m.f. measurements, rather higher concentrations of hydrochloric acid are necesary, and an extended form of the Debye–Hückel equation has to be used. The mean ionic activity coefficient is thus given by

$$\log \gamma_\pm = -A\sqrt{m} + Cm$$

where A and C are constants, A having the value of $0\cdot51$ mol$^{-\frac{1}{2}}$ kg$^{\frac{1}{2}}$ for aqueous solutions at 25°C. The value of the constant C

is immaterial in the present connection. Substituting in eqn. (6.28) and inserting the value of RT/F at 25°C

$$E_{cell} = E_{AgCl}° - 0·118 \log m + 0·06 \sqrt{m} - 0·118 \, Cm$$

or

$$E_{cell} + 0·118 \log m - 0·06 \sqrt{m} = E_{AgCl}° - 0·118 \, Cm \tag{6.29}$$

If the left-hand side of eqn. (6.29) is plotted against m, a straight line of intercept $E_{AgCl}°$ should be obtained. Extrapolation to zero molality of hydrochloric acid thus gives the standard potential of the silver–silver chloride electrode.

Most standard electrode potentials may be determined by this technique by choosing a suitable cell.

DETERMINATION OF DISSOCIATION CONSTANTS

The dissociation constants of weak acids may be determined approximately from measurements of the pH of a solution containing a mixture of the acid and one of its salts which has been formed by reaction with a strong base. Consider a solution which originally contained a moles of a weak monobasic acid, HA, to which b moles of a strong monacidic base, MOH, have been added, b being less than a. The reaction of the acid and the base may be represented as

$$HA + M^+ + OH^- \rightarrow M^+ + A^- + H_2O$$

Some of the remaining acid will be dissociated

$$HA + H_2O \rightleftharpoons H_3O^+ + A^-$$

and some of the salt will undergo hydrolysis

$$A^- + H_2O \rightleftharpoons HA + OH^-$$

The solution will contain undissociated HA molecules, H^+ ions, OH^- ions, A^- ions and M^+ ions. As the solution must be electrically neutral, the sum of all the positive charges must be equal to that of all the negative charges, and denoting the amount of each ion present in the solution as n with an appropriate subscript

$$n_{M^+} + n_{H^+} = n_{A^-} + n_{OH^-}$$

The amount of M^+ ions must be equal to that of base added, hence

$$b + n_{H^+} = n_{A^-} + n_{OH^-} \tag{6.30}$$

The A^- ions which exist in the solution must have come from the acid originally present, so that

$$a = n_{HA} + n_{A^-} \tag{6.31}$$

where n_{HA} is the amount of undissociated acid existing in the mixture. The dissociation constant of the acid is given by

$$K_a = \frac{a_{H^+} \cdot a_{A^-}}{a_{HA}}$$

$$= a_{H^+} \cdot \frac{m_{A^-}}{m_{HA}} \cdot \frac{\gamma_{A^-}}{\gamma_{HA}} \tag{6.32}$$

Suppose now that the weight of solvent in the solution is w kg, then $n_{A^-} = w m_{A^-}$ and $n_{HA} = w m_{HA}$. Multiplying the numerator and denominator of eqn. (6.32) by w gives

$$K_a = a_{H^+} \cdot \frac{w m_{A^-}}{w m_{HA}} \cdot \frac{\gamma_{A^-}}{\gamma_{HA}}$$

$$= a_{H^+} \cdot \frac{n_{A^-}}{n_{HA}} \cdot \frac{\gamma_{A^-}}{\gamma_{HA}} \tag{6.33}$$

From eqn. (6.30)

$$n_{A^-} = b + n_{H^+} - n_{OH}$$

and from eqn. (6.31)

$$n_{HA} = a - n_{A^-}$$

$$= a - b - (n_{H^+} - n_{OH^-})$$

Substituting in eqn. (6.33)

$$K_a = a_{H^+} \cdot \frac{b + n_{H^+} - n_{OH^-}}{a - b - (n_{H^+} - n_{OH^-})} \cdot \frac{\gamma_{A^-}}{\gamma_{HA}}$$

or

$$a_{H^+} = K_a \cdot \frac{a - b - (n_{H^+} - n_{OH^-})}{b + n_{H^+} - n_{OH^-}} \cdot \frac{\gamma_{HA}}{\gamma_{A^-}}$$

(6.34)

Remembering that $n_{H^+} = wm_{H^+}$ and $n_{OH^-} = wm_{OH^-}$, eqn. (6.34) may be written

$$a_{H^+} = K_a \frac{a - b - w(m_{H^+} - m_{OH^-})}{b + w(m_{H^+} - m_{OH^-})} \cdot \frac{\gamma_{HA}}{\gamma_{A^-}}$$

(6.35)

Also, as $m_{H^+} \cdot m_{OH^-} = 10^{-14}$, the ionic product for water, it may be understood that if m_{H^+} is between 10^{-4} and 10^{-10}, the term $(m_{H^+} - m_{OH^-})$ becomes negligibly small and eqn. (6.35) reduces to

$$a_{H^+} = K_a \frac{a - b}{b} \cdot \frac{\gamma_{HA}}{\gamma_{A^-}}$$

or, taking logarithms

$$pH = pK_a + \log \frac{b}{a - b} + \log \frac{\gamma_{A^-}}{\gamma_{HA}}$$

Moreover, if the solution is sufficiently dilute for the activity coefficient terms to be neglected

$$pH = pK_a + \log \frac{b}{a - b}$$

(6.36)

Equation (6.36) is identical with the simple form of the Henderson equation, (3.27), which is applicable to solutions having pH values between 4 and 10. If the amount of base added is equal to

181

half the amount of acid originally present, i.e. $b = a/2$, substitution in eqn. (6.36) gives the result

$$pH = pK_a$$

The pK_a of a weak acid is thus approximately equal to the pH of a solution of the acid to which half the equivalent amount of a strong base has been added.

If a weak acid is titrated potentiometrically with a strong base using, say, a glass electrode and a suitable reference electrode, a graph of pH against the amount of base added may be plotted. The amount of base required for neutralisation of the acid may be read from the graph together with the pH of the system when half the amount of base has been added. If this pH lies between 4 and 10, it will be approximately equal to the pK_a of the acid.

Dissociation constants of acids may be accurately determined from measurements of the e.m.f. of cells without liquid junctions such as

$$Pt; H_2/HA(m_1), NaA(m_2), NaCl(m_3), AgCl(s)/Ag$$

The e.m.f. of the cell is given by

$$E_{cell} = E_{AgCl} - E_{H_2}$$

The potentials of the silver–silver chloride electrode and the hydrogen electrodes are given by

$$E_{AgCl} = E_{AgCl}^{\circ} - \frac{RT}{F} \ln a_{Cl^-}$$

and

$$E_{H_2} = \frac{RT}{F} \ln a_{H^+}$$

assuming that the partial pressure of the hydrogen gas is 1 atm. Thus

$$E_{cell} = E_{AgCl}^{\circ} - \frac{RT}{F} \ln a_{H^+} . a_{Cl^-}$$

182

or

$$\frac{F(E_{cell} - E_{AgCl}^{\ominus})}{RT} = - \ln a_{H^+} \cdot a_{Cl^-} \qquad (6.37)$$

From eqn. (6.32)

$$a_{H^+} = K_a \cdot \frac{m_{HA}}{m_{A^-}} \cdot \frac{\gamma_{HA}}{\gamma_{A^-}}$$

and putting $a_{Cl^-} = m_{Cl^-} \gamma_{Cl^-}$ and substituting in eqn. (6.37)

$$\frac{F(E_{cell} - E_{AgCl}^{\ominus})}{RT} = - \ln K_a \cdot \frac{m_{HA}}{m_{A^-}} \cdot \frac{\gamma_{HA}}{\gamma_{A^-}} \cdot \gamma_{Cl^-} m_{Cl^-}$$

or

$$\frac{F(E_{cell} - E_{AgCl}^{\ominus})}{RT} = - \ln \frac{m_{HA} m_{Cl^-}}{m_{A^-}} - \ln \frac{\gamma_{HA} \gamma_{Cl^-}}{\gamma_{A^-}} - \ln K_a$$

or

$$\frac{F(E_{cell} - E_{AgCl}^{\ominus})}{2 \cdot 303 RT} + \log \frac{m_{HA} m_{Cl^-}}{m_{A^-}} = - \log \frac{\gamma_{HA} \gamma_{Cl^-}}{\gamma_{A^-}} - \log K_a$$
$$(6.38)$$

The e.m.f. of the cell is measured at various values of m_1, m_2 and m_3 and the left-hand side of eqn. (6.38) is plotted against the ionic strength of the solution. At zero ionic strength, the activity coefficients are equal to unity and the right-hand side of eqn. (6.38) is equal to $- \log K_a$. Thus the intercept of the graph of the left-hand side of the equation against ionic strength gives $- \log K_a$.

The molality terms in eqn. (6.38) are calculated in the following way. The sodium chloride is considered to be completely dissociated, hence

$$m_{Cl^-} = m_3$$

The acid HA will be partly dissociated into H^+ ions and A^- ions. The molality of the hydrogen ions so produced is m_{H^+}, and the stoichiometric molality of the acid is m_1. The molality of the undissociated acid molecules is thus

$$m_{HA} = m_1 - m_{H^+}$$

The A^- ions in solution are partly provided by the salt NaA and partly by the dissociation of the acid, hence

$$m_{A^-} = m_2 + m_{H^+}$$

It is still necessary to know m_{H^+}, and if the dissociation constant of the acid is not more than 10^{-4}, a sufficiently accurate value of m_{H^+} may be calculated from an approximate value of K_a. Putting

$$K_a \approx \frac{m_{H^+} m_{A^-}}{m_{HA}}$$

with $m_{A^-} \approx m_2$ and $m_{HA} \approx m_1$

$$m_{H^+} \approx K_a . \frac{m_1}{m_2}$$

In practice, the ratio of m_1 to m_2 is usually kept constant and the ionic strength of the solution is varied by using different concentrations of sodium chloride. The method as described is restricted to acids with K_a values of 10^{-4} to 10^{-5}. With stronger acids, the calculation of m_{H^+} is more complicated, and with weaker acids, allowance has to be made for the effect of hydrolysis on m_{H^+}. The method gives very good results.

DETERMINATION OF THE IONIC PRODUCT OF WATER

The ionic product of water may be determined approximately by using cells with liquid junctions. The cell

$$Pt; H_2/KOH(0 \cdot 01 \text{ mol dm}^{-3})//HCl(0 \cdot 01 \text{ mol dm}^{-3})/H_2; Pt$$

is a concentration cell where the e.m.f. depends upon the difference between the activities of the hydrogen ions in the electrode solutions, if the hydrogen gas has the same partial pressure at each electrode. Assuming that the liquid junction potential has been completely eliminated, the e.m.f. of the cell is given by

$$E_{cell} = \frac{RT}{F} \ln \frac{a''_{H^+}}{a'_{H^+}} \tag{6.39}$$

where a''_{H^+} and a'_{H^+} are the hydrogen ion activities in the hydrochloric acid and potassium hydroxide solutions, respectively. Referring to the latter, the ionic product for water may be written

$$K_w = a'_{H^+} \times a'_{OH^-}$$

where a'_{OH^-} is the hydroxide ion activity in the alkaline solution. Substituting for a'_{H^+} in eqn. (6.39)

$$E_{cell} = \frac{RT}{F} \ln \frac{a''_{H^+} \cdot a'_{OH^-}}{K_w}$$

Making the assumption that the ionic activities are equal to the mean ionic activities of the solutions

$$E_{cell} = \frac{RT}{F} \ln \frac{(a_\pm)''(a_\pm)'}{K_w}$$

where $(a_\pm)''$ and $(a_\pm)'$ are the mean ionic activities of the hydrochloric acid and the potassium hydroxide solutions, respectively. A measurement of the e.m.f. of the cell would thus give an approximate value of K_w.

Accurate values of K_w may be obtained from cells without liquid junctions. An example of a suitable cell is provided by

$$Pt; H_2/KOH\ (m_1),\ KCl\ (m_2),\ AgCl(s)/Ag$$

where the partial pressure of hydrogen gas is 1 atm. The e.m.f. of the cell is given by

$$E_{cell} = E_{AgCl} - E_{H_2}$$

The potentials of the individual electrodes are given as

$$E_{AgCl} = E_{AgCl}^{\bullet} - \frac{RT}{F} \ln a_{Cl^-}$$

185

and

$$E_{H_2} = \frac{RT}{F} \ln a_{H^+}$$

Hence

$$E_{cell} = E_{AgCl}^\circ - \frac{RT}{F} \ln a_{H^+} . a_{Cl^-}$$

Since

$$K_w = a_{H^+} . a_{OH^-}$$

$$E_{cell} = E_{AgCl}^\circ - \frac{RT}{F} \ln K_w - \frac{RT}{F} \ln \frac{a_{Cl^-}}{a_{OH^-}}$$

$$= E_{AgCl}^\circ - \frac{RT}{F} \ln K_w - \frac{RT}{F} \ln \frac{m_{Cl^-}}{m_{OH^-}} - \frac{RT}{F} \ln \frac{\gamma_{Cl^-}}{\gamma_{OH^-}}$$

Rearranging

$$E_{cell} - E_{AgCl}^\circ + \frac{RT}{F} \ln \frac{m_{Cl^-}}{m_{OH^-}} = - \frac{RT}{F} \ln K_w - \frac{RT}{F} \ln \frac{\gamma_{Cl^-}}{\gamma_{OH^-}}$$

Assuming that the potassium hydroxide and the potassium chloride are completely dissociated, m_{OH^-} may be put equal to m_1 and m_{Cl^-} equal to m_2. Making these substitutions and converting to common logarithms

$$\frac{F(E_{cell} - E_{AgCl}^\circ)}{2 \cdot 303 RT} + \log \frac{m_2}{m_1} = - \log K_w - \log \frac{\gamma_{Cl^-}}{\gamma_{OH^-}}$$

(6.40)

At zero concentration, the activity coefficients are equal to unity, and under these conditions the right-hand side of eqn. (6.40) is equal to $- \log K_w$. If the left-hand side of the equation is plotted against ionic strength, therefore, the intercept will give $- \log K_w$.

7

ELECTROLYSIS

ELECTRODE PROCESSES

CONSIDER an electrode of a metal, M, partially immersed in a solution of metal ions, M^+. It has been explained in Chapter 4 that an equilibrium between two opposing reactions is established:

$$M \rightleftharpoons M^+ + e$$

and the electrode adopts a potential depending on the position of the equilibrium. The further the equilibrium lies to the right, the more negative will be the electrode potential as the electrons accumulate in the metal. As the system is in equilibrium, the potential is said to be the reversible potential of the electrode.

If the electrode were included in a circuit and connected to a source of potential which was infinitesimally more positive than the reversible potential of the electrode, electrons would flow from the electrode into the circuit. This loss of electrons would disturb the above equilibrium, and a net reaction would occur in an attempt to restore the equilibrium by supplying more electrons. The reaction is obviously

$$M \rightarrow M^+ + e$$

The metal electrode dissolves to form metal cations which move away from the electrode into the solution. The electrode is thus functioning as an anode. If the external potential which is applied to the electrode is only infinitesimally more positive than the reversible electrode potential, the rate of loss of electrons from the electrode will be small and the electrode reaction easily keeps pace with the loss of electrons. Under these conditions, the electrode will maintain its reversible potential and is said to operate reversibly as an anode.

If, however, the applied potential is much more positive than the reversible potential of the electrode, electrons will leave the electrode at a greater rate than in the reversible operation. If there is a slow stage in the dissolution of the metal, the rate of loss of electrons will be greater than that of production by the electrode reaction. Under these conditions, the potential of the electrode would become more positive than the reversible potential and the electrode is said to be *polarised*. Any electrode operating at a potential other than its reversible potential is said to be polarised and to be operating *irreversibly*.

As has been shown, the irreversible potential of an anode is more positive than its reversible potential and the relationship may be expressed as

$$E_A' > E_{rev} \qquad (7.1)$$

where E_A' is the potential of an anode operating irreversibly and E_{rev} is its reversible potential.

When the applied potential is infinitesimally more negative than the reversible potential of the electrode, electrons will flow to the electrode from the external source. This will disturb the equilibrium and a net reaction occurs which consumes the excess electrons. This reaction must be

$$M^+ + e \rightarrow M$$

Cations are moving through the solution towards the electrode to accept electrons and deposit as metal atoms. The electrode is thus functioning as a cathode. If the applied potential is only infinitesimally more negative than the reversible electrode potential, the current flowing will be infinitesimal and the electrode reaction can keep pace with the influx of electrons. The electrode thus maintains its reversible potential. If the applied potential is appreciably more negative, electrons arrive at the electrode at an appreciable rate. If there is a slow stage in the deposition of the metal, electrons will arrive more rapidly than they can be removed and the potential of the electrode becomes more negative than the reversible potential. The electrode is thus polarised and operating irreversibly. The irreversible potential of a cathode is therefore more negative than the reversible potential. The relationship may be expressed as

$$E_C' < E_{rev} \tag{7.2}$$

where E_C' is the potential of a cathode operating irreversibly and E_{rev} is its reversible potential.

It will be understood from the above that *polarisation always makes anodes more positive and cathodes more negative than their reversible potentials*. Moreover, for an electrode to behave as a cathode it must have applied to it a potential more negative than its reversible potential and conversely for anodic behaviour.

The amount by which the potential of a working electrode deviates from its reversible potential can be expressed by the overpotential, η, which is defined by

$$E' = E + \eta \tag{7.3}$$

where E' is the working potential of the electrode and E is its reversible or equilibrium potential. It will be appreciated from the relationships (7.1) and (7.2) that for anodes η will be a positive quantity and for cathodes it will be negative.

As has been pointed out above, polarisation results from a slow stage in an electrode process. The discharge of ions at an electrode involves three main stages:

(1) transport of ions to the electrode surface,

(2) discharge of the ions to form atoms,

(3) conversion of the atoms to the normal stable form. In the case of a metal, this involves fitting the atom into an appropriate place in the lattice or, in the case of a gas, the combination of atoms to form molecules. The dissolution of a metal or a gas to form ions is merely the converse of the processes described.

Any of the three processes may be the slow stage which determines the rate at which electrons are transferred and hence the current. The overpotential which arises from items (2) and (3) is called *activation overpotential* and that arising from item (1) is called *concentration overpotential*. Activation overpotential is important in the evolution of gases, particularly hydrogen and oxygen. In the deposition or dissolution of metals however, the vast majority of the overpotential may be attributed to concentration overpotential, with the exception of a few cases such as iron, cobalt and nickel where activation overpotential plays an important part.

189

ELEMENTARY ELECTROCHEMISTRY

Although a metal electrode has been considered as an example, the arguments and conclusions above apply equally to any electrode system in any situation. It should be noted in passing that anodic reactions are oxidation reactions and cathodic reactions are reduction reactions.

It will also be clear from the above discussion that the kinetics of the electrode reactions are very important when considering irreversible electrode processes. Up to now in Chapters 4 and 6 only reversible situations have been considered in relation to electrode potentials and, in consequence, we have been able to treat the problems thermodynamically. We must now consider electrode potentials from a kinetic point of view and it will be of value to consider reversible potentials from this standpoint to afford a comparison with the thermodynamic treatment. We shall therefore consider the forward and backward reactions involved in the equilibrium established at any electrode which is at its reversible electrode potential. In doing so we have to consider the transfer of charge across the electrode/solution interface and it will be useful to have some idea of the structure of this interface before proceeding to the kinetics.

THE ELECTRICAL DOUBLE LAYER

Between a metal and an electrolyte solution there is a potential difference which is due to an unequal distribution of charges across the interface. The potential difference may arise as a result of charge transfer reactions such as occur when an electrode adopts its reversible potential or it may be due to the application of a potential to the electrode from an external source. The nature of the charge distribution is of interest in connection with the mechanisms of electrode reactions. The first and simplest concept of the charge distribution is due to Helmholtz who suggested that the potential difference lay between two layers of electrical charges of opposite sign. This was the origin of the term *electrical double layer* used to denote the array of charged particles existing at the interface. In the case of a metal/electrolyte interface one layer of charges will be in the metal. Suppose that the metal is negatively charged with respect to the solution, then on the solution side of the interface Helmholtz suggested that there was a layer of positively charged ions immediately adjacent to the metal surface. Such a system is similar to a parallel plate condenser and the double layer should have a capacitance just like a condenser. Measurements of the capacitance of the electrical double layer and its dependence on parameters such as electrolyte concentration showed that the Helmholtz model was inadequate and Gouy and Chapman suggested that the structure of the double layer on the solution side was not rigid but *diffuse* due to thermal agitation of the ions in the solution and the electrical forces between these ions. The theory of the completely diffuse double layer did not give good agreement with experimentally determined double layer capacitances and Stern suggested that the electrical double layer was a combination of a rigid layer in which some substances were adsorbed on the metal surface and a diffuse layer. The modern qualitative picture of the electrical double layer at a metal/solution interface is due to Grahame who considered the question of solvation of the ions which were adsorbed on the electrode surface.

Firstly, the metallic phase has a net electrical charge due to an excess or deficit of electrons. This charge is confined to so thin a layer in the surface of the metal that it may be considered to be two dimensional. The Helmholtz layer on the solution side of the interface contains solvent molecules and sometimes other neutral molecules adsorbed on the metal surface. If these molecules are dipoles (as when the solvent is water), they will be oriented to some extent due to the charge on the metal. In addition, in most electrolyte solutions there are some ions (a fractional monolayer, usually of anions) adsorbed on the metal surface. The forces involved in this adsorption are not clear but before an ion can be adsorbed in this way it must be unsolvated, at least on the side in contact with the electrode. Ions adsorbed in this way are said to be *specifically adsorbed* and the plane through the electrical centres of these ions is called the *inner Helmholtz plane*. Further away from the electrode there is a layer of solvated ions. These are non-specifically adsorbed ions and the plane through the centres of such ions closest to the electrode is called the *outer Helmholtz plane* which marks the boundary of the Helmholtz layer and the beginning of the diffuse layer which contains other non-specifically adsorbed ions. This qualitative picture of the electrical double layer is illustrated in *Figure 30* where it is assumed that the metal is negatively charged.

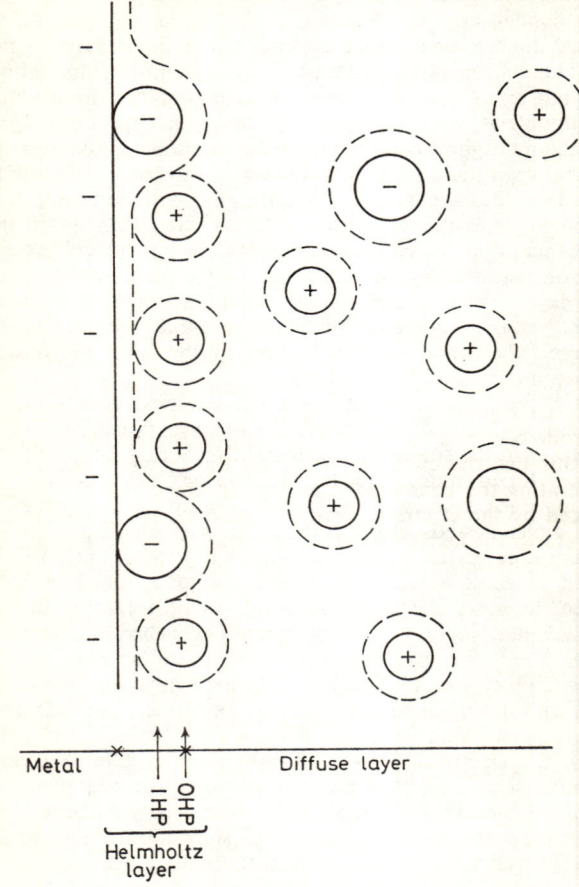

Figure 30. The electrical double layer

ELECTROLYSIS

As mentioned before, the electrical double layer is of interest because the measured values of the rate constants of electrode reactions and other kinetic parameters can be strongly dependent on the structure of the double layer. Neglect of double layer structure can lead to erroneous mechanisms. Although such considerations are outside the scope of this book it is as well to know of the possible effects.

THE KINETIC TREATMENT OF
REVERSIBLE ELECTRODE POTENTIALS

When an electrode is at its reversible potential, there exists an equilibrium between the oxidised and reduced forms of the system. In general the equilibrium may be represented

$$Ox + ze \rightleftharpoons Red$$

It has been explained in Chapter 4, that initially (before any potential difference between the electrode and the solution is developed), the forward and backward reactions do not proceed at the same rate. If the forward reaction is faster initially, the electrode becomes positively charged with respect to the solution and this has the effect of retarding the forward reaction and accelerating the backward reaction. In every case the potential adopted by the electrode always accelerates the slower reaction and retards the faster reaction until equilibrium is achieved when the speeds of both reactions are equal. At this point the potential difference between the electrode and the solution ceases to increase and remains at a steady value. We must now examine the kinetic aspects of this process.

Suppose that in the absence of any potential difference the velocities of the forward and backward reactions are v_1^0 and v_2^0 respectively, the superscript indicating initial conditions. These velocities can be expressed in terms of velocity constants and the activities of the reactants. Thus

$$v_1^0 = k_1^0 a_O \quad \text{and} \quad v_2^0 = k_2^0 a_R \tag{7.4}$$

where k_1^0 and k_2^0 are the velocity constants in the absence of an potential and a_O and a_R are the activities of the oxidised and reduced forms respectively. From the theory of absolute reaction rates the velocity constants can be expressed in terms of the standard free energy of activation of the reactions

$$k_1^0 = \frac{kT}{h} \exp\left(-\frac{\Delta G_1^{\neq}}{RT}\right) \quad k_2^0 = \frac{kT}{h} \exp\left(-\frac{\Delta G_2^{\neq}}{RT}\right) \quad (7.5)$$

where $\Delta G_1^{\neq}$ and $\Delta G_2^{\neq}$ are the standard free energies of activation of the forward and backward reactions respectively. The position can be illustrated by a reaction co-ordinate diagram as shown in *Figure 31*.

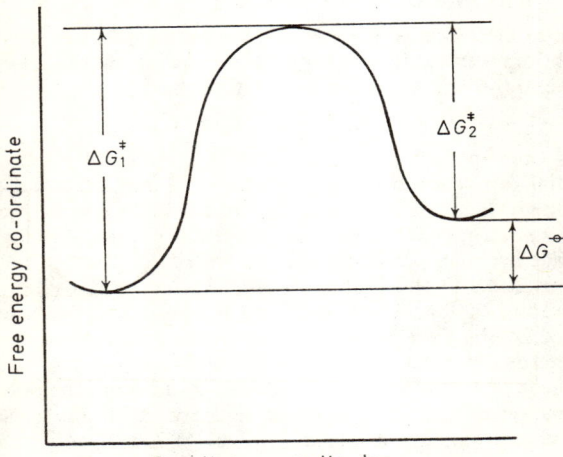

Figure 31. Reaction co-ordinate diagram

If $v_1^0 \neq v_2^0$, the electrode will adopt a potential E which accelerates the slower reaction and retards the faster reaction. As the change in the activities of the reactants will be negligible, this effect must arise from a change in the velocity constants which, in turn, can only be affected by a change in the heights of the energy barriers for the reactions. Suppose that initially $v_1^0 < v_2^0$ so that the potential accelerates the forward reaction and retards the backward reaction. Suppose that the activation energy of the forward reaction is decreased by an amount $\Delta G_1{}^E$ and that the activation energy of the backward reaction is increased by an amount $\Delta G_2{}^E$. The situation will be as represented in *Figure 32*. As a result of the change in the heights of the energy

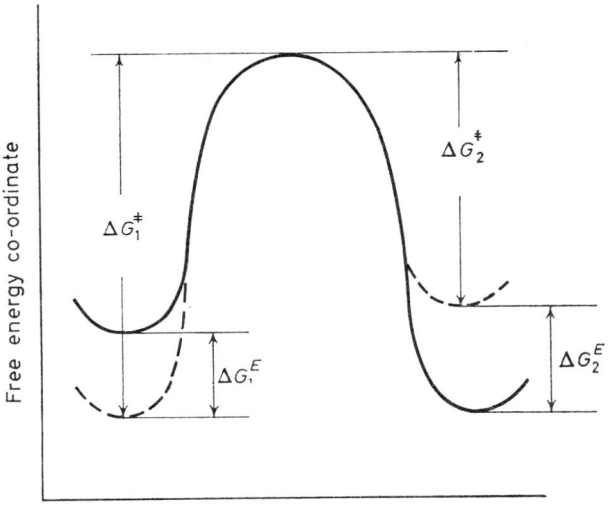

Reaction co-ordinate

Figure 32. Reaction co-ordinate diagram

barriers due to the potential E, the velocities of the forward and backward reactions will change. Let these now be represented by v_1 and v_2 respectively, k_1 and k_2 being the corresponding rate constants. Now

$$v_1 = k_1 a_O \quad \text{and} \quad v_2 = k_2 a_R \tag{7.6}$$

where

$$k_1 = \frac{kT}{h} \exp\left(-\frac{\Delta G_1{}^{\neq} - \Delta G_1{}^E}{RT}\right)$$

195

$$k_2 = \frac{kT}{h} \exp\left(-\frac{\Delta G_2^{\neq} - \Delta G_2^{E}}{RT}\right) \tag{7.7}$$

Suppose that a fraction α of the potential E went towards facilitating the forward reaction whilst the remaining fraction $(1-\alpha)$ retarded the backward reaction, then

$$\Delta G_1^{E} = -\alpha z F E \quad \text{and} \quad \Delta G_2^{E} = -(1-\alpha)z F E$$

Substituting in eqn. (7.5)

$$k_1 = \frac{kT}{h} \exp\left(-\frac{\Delta G_1^{\neq} + \alpha z F E}{RT}\right)$$

$$k_2 = \frac{kT}{h} \exp\left(-\frac{\Delta G_2^{\neq} - (1-\alpha)z F E}{RT}\right)$$

or

$$k_1 = \frac{kT}{h} \exp\left(-\frac{\Delta G_1^{\neq}}{RT}\right) \exp\left(-\frac{\alpha z F E}{RT}\right)$$

$$k_2 = \frac{kT}{h} \exp\left(-\frac{\Delta G_2^{\neq}}{RT}\right) \exp\left(\frac{(1-\alpha)z F E}{RT}\right) \tag{7.8}$$

Comparing this result with eqn. (7.5) it may be seen that

$$k_1 = k_1^0 \exp\left(-\frac{\alpha z F E}{RT}\right)$$

$$k_2 = k_2^0 \exp\left(\frac{(1-\alpha)z F E}{RT}\right) \tag{7.9}$$

These equations relate the rate constant of an electrode reaction in the presence of a potential E to that in the absence of any potential.

The actual velocities of the reactions are given by eqns. (7.6) as

$$v_1 = a_o k_1^0 \exp\left(-\frac{\alpha z F E}{RT}\right)$$

$$v_2 = a_R k_2^0 \exp\left(\frac{(1-\alpha)z F E}{RT}\right) \tag{7.10}$$

At equilibrium, E will be the reversible potential of the electrode and $v_1 = v_2$, hence

$$a_O k_1{}^0 \exp\left(-\frac{\alpha z F E}{RT}\right) = a_R k_2{}^0 \exp\left(\frac{(1-\alpha)z F E}{RT}\right)$$

Taking logarithms

$$\ln a_O + \ln k_1{}^0 - \frac{\alpha z F E}{RT} = \ln a_R + \ln k_2{}^0 + \frac{(1-\alpha)z F E}{RT}$$

Rearranging

$$\frac{zFE}{RT} = \ln \frac{a_O}{a_R} + \ln \frac{k_1{}^0}{k_2{}^0}$$

or

$$E = \frac{RT}{zF} \ln \frac{k_1{}^0}{k_2{}^0} + \frac{RT}{zF} \ln \frac{a_O}{a_R} \tag{7.11}$$

Equations (7.5) give $k_1{}^0$ and $k_2{}^0$ in terms of free energies of activation. Using these expressions

$$\ln \frac{k_1{}^0}{k_2{}^0} = \ln \left[\exp\left(-\frac{\Delta G_1{}^{\neq}}{RT}\right) \exp\left(\frac{\Delta G_2{}^{\neq}}{RT}\right) \right]$$

or

$$\ln \frac{k_1{}^0}{k_2{}^0} = \frac{\Delta G_2{}^{\neq} - \Delta G_1{}^{\neq}}{RT}$$

From *Figure 31*, it can be seen that $(\Delta G_2{}^{\neq} - \Delta G_1{}^{\neq}) = -\Delta G^\circ$, hence

$$\ln \frac{k_1{}^0}{k_2{}^0} = -\frac{\Delta G^\circ}{RT}$$

Remembering that $\Delta G^\circ = -zFE^\circ$

$$\ln \frac{k_1{}^0}{k_2{}^0} = \frac{zFE^\circ}{RT}$$

Substituting in eqn. (7.11)

$$E = E^\circ + \frac{RT}{zF} \ln \frac{a_O}{a_R} \tag{7.12}$$

which is the equation deduced thermodynamically in Chapter 4 for reversible electrode potentials.

Before leaving the equilibrium case a further point should be mentioned. The forward and backward reactions are heterogeneous reactions in so far as they occur at the surface of the electrode. The velocities are thus expressed in units of (amount of substance) (time)$^{-1}$ (area)$^{-1}$. As charge is transferred across the double layer in the course of these reactions, each reaction corresponds to an electric current and the velocities of the reactions can be expressed in terms of current densities by multiplying the velocity by zF. For example, suppose $v_1 = 10^{-8}$ mol cm^{-2} s^{-1} and assume that $z = 1$, then

$$zFv_1 = 1 \times 96\,500\,\text{C mol}^{-1} \times 10^{-8}\,\text{mol cm}^{-2}\,\text{s}^{-1}$$
$$= 9{\cdot}65 \times 10^{-4}\,\text{C cm}^{-2}\,\text{s}^{-1}$$
$$= 9{\cdot}65 \times 10^{-4}\,\text{A cm}^{-2}$$

Furthermore, the forward reaction is a reduction reaction and hence a cathodic reaction, the backward reaction being an oxidation reaction and hence an anodic reaction. The speed of the forward reaction can thus be expressed in terms of a cathodic current density j_c, and that of the backward reaction can be expressed as an anodic current density j_a.

$$j_c = zFa_O k_1{}^0 \exp\left(-\frac{\alpha zFE}{RT}\right)$$
$$j_a = zFa_R k_2{}^0 \exp\left(\frac{(1-\alpha)zFE}{RT}\right) \qquad (7.13)$$

As, at equilibrium, $v_1 = v_2$, j_c must be equal to j_a under these conditions. This common value of j_c and j_a is denoted j_o and called the *exchange current density*. Thus

$$j_o = zFa_O k_1{}^0 \exp\left(-\frac{\alpha zFE}{RT}\right) = zFa_R k_2{}^0 \exp\left(\frac{(1-\alpha)zFE}{RT}\right)$$
$$(7.14)$$

where E is the reversible potential of the electrode.

An electrode at equilibrium thus behaves simultaneously as an anode and a cathode but the anodic current is exactly equal to the cathodic current so that the net current is zero.

It should also be pointed out that the exchange current density defined by eqns. (7.14) might better be called an apparent exchange current density because it will depend to some extent upon the structure of the double layer. The same considerations apply to the parameter α which is called the *transfer coefficient*. These points become important in the elucidation of the mechanisms of electrode reactions but the detailed treatment is outside the scope of this book.

ACTIVATION OVERPOTENTIAL

Activation overpotential arises when the actual electrode process is the rate determining step rather than transport of the ion to the electrode surface which is frequently called a mass transport effect. The kinetic approach used in the previous section can be employed in the consideration of activation overpotential and for the sake of simplicity we shall assume that the electrode process is not dependent upon mass transport effects. We shall further assume that there is no specific adsorption of reactants or products on the electrode.

Suppose that a potential E', other than the reversible potential E, is impressed on an electrode from an external source. This applied potential will alter the rates of the forward and backward reactions causing one to be accelerated and the other to be retarded. If $E' > E$, the anodic reaction will be accelerated and the cathodic reaction will be retarded. This means that $j_a > j_c$ and the electrode will pass a net current which will be anodic. Conversely, if $E' < E$, the cathodic reaction proceeds faster than the anodic reaction and the net behaviour of the electrode will be cathodic.

Suppose that in the presence of the potential E' the velocities of the cathodic and anodic reactions are v_1' and v_2' respectively, the corresponding rate constants being k_1' and k_2'. Then

$$v_1' = k_1' a_O \quad \text{and} \quad v_2' = k_2' a_R \qquad (7.15)$$

The rate constants k_1' and k_2' are related to k_1^0 and k_2^0 by equations similar to eqns. (7.9).

$$k_1' = k_1^0 \exp\left(-\frac{\alpha z F E'}{RT}\right)$$

$$k_2' = k_2^0 \exp\left(\frac{(1-\alpha)zFE'}{RT}\right) \tag{7.16}$$

where α is the fraction of the potential E' favouring the cathodic reaction and a fraction $(1-\alpha)$ retards the anodic reaction. From eqn. (7.3)

$$E' = E + \eta \tag{7.3}$$

and substituting for E' in eqns. (7.16)

$$k_1' = k_1^0 \exp\left(-\frac{\alpha z F (E+\eta)}{RT}\right)$$

$$k_2' = k_2^0 \exp\left(\frac{(1-\alpha)zF(E+\eta)}{RT}\right)$$

or

$$k_1' = k_1^0 \exp\left(-\frac{\alpha z F E}{RT}\right) \exp\left(-\frac{\alpha z F \eta}{RT}\right)$$

$$k_2' = k_2^0 \exp\left(\frac{(1-\alpha)zFE}{RT}\right) \exp\left(\frac{(1-\alpha)zF\eta}{RT}\right) \tag{7.17}$$

Substituting from eqns. (7.17) into eqns. (7.15), the velocities of the reactions are given by

$$v_1' = a_O k_1^0 \exp\left(-\frac{\alpha z F E}{RT}\right) \exp\left(-\frac{\alpha z F \eta}{RT}\right)$$

$$v_2' = a_R k_2^0 \exp\left(\frac{(1-\alpha)zFE}{RT}\right) \exp\left(\frac{(1-\alpha)zF\eta}{RT}\right)$$

Expressing these velocities in terms of current densities

$$j_c = z F a_O k_1^0 \exp\left(-\frac{\alpha z F E}{RT}\right) \exp\left(-\frac{\alpha z F \eta}{RT}\right)$$

$$j_a = zFa_Rk_2{}^0 \exp \left(\frac{(1 - \alpha)zFE}{RT} \right) \exp \left(\frac{(1 - \alpha)zF\eta}{RT} \right) \quad (7.18)$$

Comparison with eqn. (7.14) shows that eqns. (7.18) may be written

$$j_c = j_o \exp \left(- \frac{\alpha zF\eta}{RT} \right) \quad j_a = j_o \exp \left(\frac{(1 - \alpha)zF\eta}{RT} \right) \quad (7.19)$$

From eqns. (7.19) it can be seen that if η is negative, $j_c > j_a$ and the electrode behaves as a cathode. Conversely, if η is positive, $j_c < j_a$ and the net behaviour of the electrode is anodic. It will be appreciated that at any electrode an anodic and a cathodic current may be considered to pass, the net behaviour simply depending on the relative magnitudes of these two currents.

It is the usual convention to write the net current density j, as

$$j = j_c - j_a \quad (7.20)$$

Thus if $j_c > j_a$. j is positive so that cathodic currents are, by convention, considered to be positive. Alternatively, if $j_c < j_a$, j will be negative so that anodic currents are conventionally allotted a minus sign.

The variation of the individual current densities j_c and j_a with overpotential may be represented graphically in a plot of current density against overpotential. In such plots it is usual to represent j_a as a negative quantity and to plot negative values of η to the right of the origin as illustrated in *Figure 33*. Considering j_c in eqns. (7.19), when $\eta = 0$, $j_c = j_o$. Further, as $\eta \rightarrow + \infty$, $j_c \rightarrow 0$ and as $\eta \rightarrow - \infty$, $j_c \rightarrow \infty$. For j_a in eqn. (7.19), when $\eta = 0$, $j_a = j_o$. Furthermore as $\eta \rightarrow - \infty$, $j_a \rightarrow 0$ and as $\eta \rightarrow + \infty$, $j_a \rightarrow \infty$. A plot of eqns. (7.19) is shown in *Figure 33* where anodic current densities are represented as negative. The net current density j is the sum of j_c and $-j_a$ and is shown by the graph which goes through the origin. Thus, when $\eta = 0$, $j = 0$. When η is negative, j is positive indicating a net cathodic current and when η is positive, j is negative indicating a net anodic current.

The equation for the graph of the net current density is obtained by substituting from eqns. (7.19) into eqn. (7.20). Thus

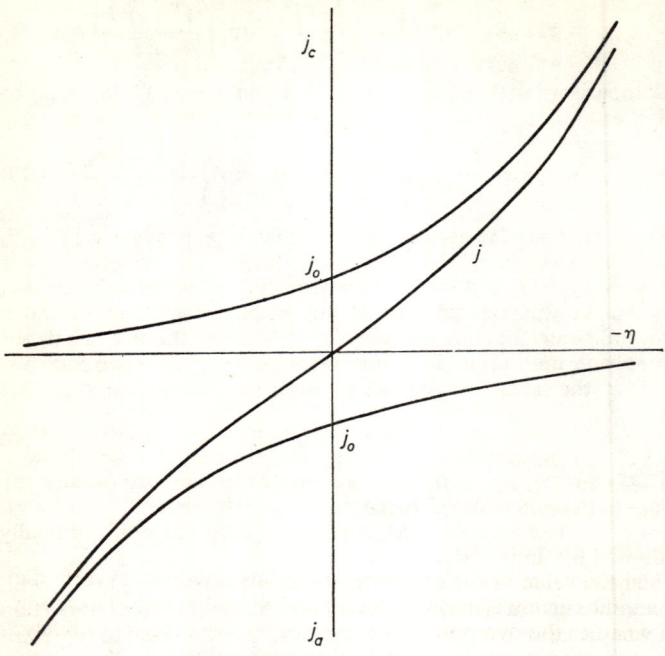

Figure 33. Current—overpotential curves

$$j = j_o \left[\exp \left(- \frac{\alpha z F \eta}{RT} \right) - \exp \left(\frac{(1 - \alpha) z F \eta}{RT} \right) \right] \quad (7.21)$$

This equation relates the current density at an electrode to the overpotential. It can be simplified in two extreme cases.

(a) Low overpotential

Exponential terms can be expressed as a power series

$$e^x = 1 + x + \frac{x^2}{2!} + \frac{x^3}{3!} + \dots$$

202

If x is small, $e^x \approx 1 + x$. For cases of low overpotential this approximation may be made for the exponential terms in eqn. (7.21) when

$$j = j_o \left[1 - \frac{\alpha z F \eta}{RT} - \left(1 + \frac{(1 - \alpha) z F \eta}{RT} \right) \right]$$

or

$$j = - j_o \frac{z F \eta}{RT}$$

Rearranging

$$\eta = - \frac{RT}{zF} \frac{j}{j_o} \tag{7.22}$$

The negative sign indicates that if η is positive, j must be negative showing that the net current is anodic. For low overpotentials, eqn. (7.22) shows that j is directly proportional to η and it can be seen in *Figure 33* that the graph of net current density is linear in the region of the origin.

(b) High overpotential

When the value of the overpotential is large, either the cathodic or anodic current is negligible compared with the other depending on whether the overpotential is positive or negative respectively. The two cases must thus be considered separately.

(i) Net cathodic behaviour

If the overpotential is large and negative the second term in the bracket in eqn. (7.21) is negligible compared with the first and the equation reduces to

$$j = j_o \exp \left(- \frac{\alpha z F \eta}{RT} \right)$$

Taking logarithms

$$\ln j = \ln j_o - \frac{\alpha z F \eta}{RT}$$

Rearranging and converting to common logarithms

$$\eta = \frac{2\cdot303RT}{\alpha zF} \log j_o - \frac{2\cdot303RT}{\alpha zF} \log j \qquad (7.23)$$

Putting

$$\frac{2\cdot303RT}{\alpha zF} \log j_o = a \qquad (7.24)$$

and

$$\frac{2\cdot303RT}{\alpha zF} = b \qquad (7.25)$$

$$\eta = a - b \log j \qquad (7.26)$$

Equation (7.26) relates overpotential to net cathodic current density and is known as the *Tafel equation*. It shows that at high values the overpotential is proportional to the logarithm of the current density.

(ii) *Net anodic behaviour*

In this case it is the first term in the bracket of eq. (7.21) which is negligible with respect to the second and the equation may be written

$$-j = j_o \exp \left(\frac{(1 - \alpha)zF\eta}{RT} \right)$$

In order to take logarithms of this equation the negative sign attached to the net anodic current density must be neglected and we consider the modulus of this quantity. Thus

$$\ln |j| = \ln j_o + \frac{(1 - \alpha)zF\eta}{RT}$$

Converting to common logarithms and rearranging

$$\eta = -\frac{2\cdot303RT}{(1 - \alpha)zF} \log j_o + \frac{2\cdot303RT}{(1 - \alpha)zF} \log |j|$$

Putting

$$-\frac{2 \cdot 303RT}{(1 - \alpha)zF} \log j_o = a' \qquad (7.27)$$

and

$$\frac{2 \cdot 303RT}{(1 - \alpha)zF} = b' \qquad (7.28)$$

$$\eta = a' + b' \log |j| \qquad (7.29)$$

Equation (7.29) is thus the form of the Tafel equation which is applicable to net anodic behaviour.

It can be seen from eqns. (7.24) and (7.25) and also from eqns. (7.27) and (7.28) that

$$a/b = -a'/b' = \log j_o \qquad (7.30)$$

If the overpotential of an electrode is determined for various net current densities and a graph of η against $\log j$ is plotted, the Tafel equation constants a (or a') and b (or b') may be obtained from the intercept and the slope of the graph respectively. The exchange current density of an electrode system may thus be calculated from eqn. (7.30). The transfer coefficient may be calculated from the slope of the graph.

MEASUREMENT OF THE POTENTIAL OF A WORKING ELECTRODE

The method of determining the potential of a working electrode is, in principle, the same as that for determining a reversible electrode potential but the fact that the electrode is passing a current involves some degree of modification. Suppose that the working electrode of interest is operating as a cathode. It must of course form part of a complete circuit which will include an anode (sometimes known as the auxiliary electrode). The potential of the working electrode is determined by coupling it with a reference electrode and measuring the e.m.f. of the cell so formed with a potentiometer. This principle is illustrated in *Figure 34*. The working current flows round the circuit between

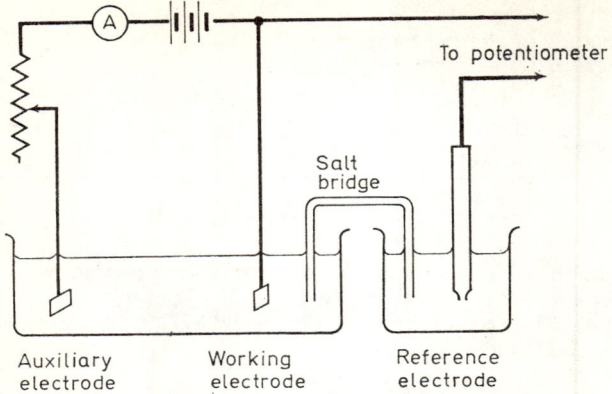

To potentiometer

Salt bridge

Auxiliary electrode

Working electrode

Reference electrode

Figure 34. Determination of the potential of a working electrode

the working electrode and the auxiliary electrode but no current flows in the potentiometer circuit which is, of course, balanced for zero current. The working current does flow through the electrolyte in the electrolysis cell however, so that between the working electrode and the end of the salt bridge there will be an IR drop through the electrolyte solution. This will be included in the e.m.f. of the cell formed by the working electrode and the reference electrode and will thus introduce an error into the measurements. This problem may be overcome in the following ways. The salt bridge may be extended to terminate in a capillary, the tip of which is placed as near the working electrode as possible. This device is known as a Luggin capillary and is illustrated in *Figure 35*. As no current flows in the bridge, the Luggin capillary has the effect of extending the reference electrode, the amount of electrolyte between the working electrode and the tip of the capillary (and hence the IR drop therein) being reduced to very small proportions.

The effectiveness of the Luggin capillary depends upon the current passing and the concentration of the electrolyte. If the electrolyte is dilute, its resistance will be high leading to a large

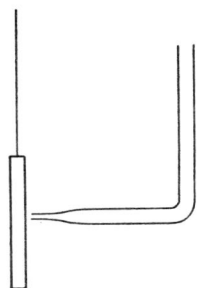

Figure 35. Luggin capillary

IR drop as will also be the case for large currents. Under these conditions the *IR* drop between the tip of the Luggin capillary and the electrode may still introduce an appreciable error. In this case the potential of the working electrode may be measured for a particular current with the tip of the capillary at several measured distances from the working electrode. The farther away the capillary is, the greater will be the *IR* error but by plotting a graph of measured overpotential against distance of the capillary from the electrode the graph may be extrapolated to zero distance to give the true overpotential as illustrated in *Figure 36* for a case of cathodic overpotential.

Another method depends on the fact that the *IR* drop through an electrolyte solution becomes zero as soon as the current is switched off whilst the overpotential decays at first linearly and then exponentially with time. The electrode potential is thus measured at several different intervals of time after the current has been interrupted. The observed overpotential is plotted against the time after interruption of the current. If the linear portion of the decay curve can be traced, it can be extrapolated to zero time to give the true overpotential. The decay of the over-potential is usually observed up to times of about 10^{-3} s so that the interruption of the current, the delay time and the potential measurement are controlled by electronic circuits.

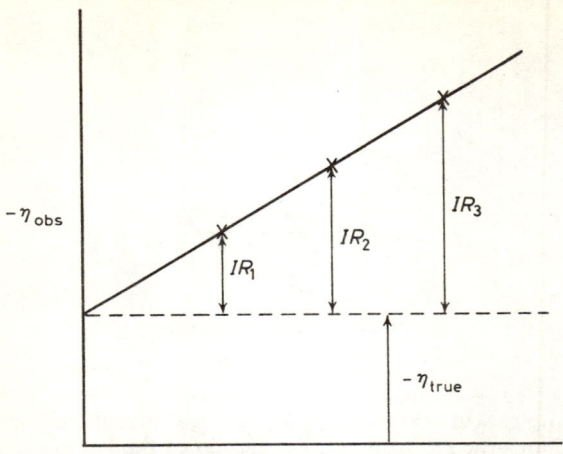

Electrode capillary distance

Figure 36. Elimination of IR error by extrapolation

THE SIGNIFICANCE OF EXCHANGE CURRENT DENSITY

The exchange current density is a measure of the speeds of the anodic and cathodic reactions occurring at an electrode when it is at its reversible potential. It will be recalled that activation polarisation results from a slow stage in the electrode reaction. The slower the electrode reaction, the more easily will polarisation develop. As the exchange current density is a measure of the speed of the electrode reaction, it should provide an indication of the liability to polarisation of a particular electrode system. An electrode with a low exchange current density will have a slow reaction and thus be more easily polarised.

These points can be appreciated from a consideration of the equations relating overpotential to current density. For low overpotentials the appropriate equation is eqn. (7.22)

$$\eta = -\frac{RT}{zF}\frac{j}{j_o} \tag{7.22}$$

From this equation it can be seen that for a given net current density j, the magnitude of the overpotential will be less, the greater is the value of j_o. In other words, the lower the exchange current density of an electrode the greater will be the polarisation and overpotential for a given current density.

For high overpotentials at a cathode eqn. (7.23) applies and may be written in the form

$$-\eta = b(\log j - \log j_o) \qquad (7.31)$$

From this relation also it may be seen that for a given net current density j, the magnitude of the cathodic overpotential, $-\eta$, will be less the larger the value of j_o.

The point may be illustrated graphically in plots of current densities against overpotential, as in *Figure 33*, but contrasting a case of low exchange current density with one of high exchange current density. This is done in *Figure 37* from which it may be seen that for a particular value of net current density the overpotential is much greater when j_o is small.

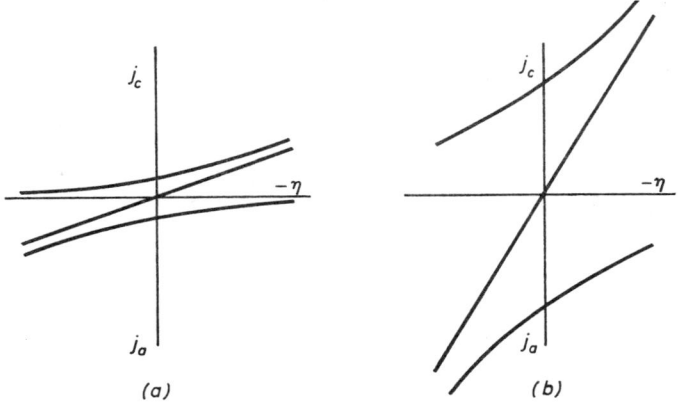

Figure 37. Current density—overpotential plots
(a) small j_o (b) large j_o

At the beginning of this Chapter it was stated that if only a small current flowed an electrode would operate reversibly and maintain its reversible potential. It will now be appreciated that if any current flows the potential of an electrode will depart from its reversible value. It would be more accurate to state that the slower an electrode reaction, the greater will be the departure from the reversible potential. Thus, all electrodes operate irreversibly and strictly they should be classed as more irreversible or less irreversible although the terms irreversible and reversible are frequently used in this context. Exchange current densities typically fall in the range 10^{-2}–10^{-16} A cm^{-2} so that an electrode with $j_o = 10^{-2}$ A cm^{-2} will pass large currents for small values of overpotential.

Another important aspect of exchange current density is its bearing on the ease with which the equilibrium potential may be established at an electrode. For the establishment of equilibrium charge must be transferred across the double layer by the anodic and cathodic reactions. If $j_o < 10^{-7}$ A cm^{-2} it is unlikely that the electrode will attain its equilibrium potential because even in conventionally pure systems there will probably exist sufficient impurities to maintain a current density greater than 10^{-7} A cm^{-2}. In this case, the potential adopted by the electrode will be due to the impurities rather than the system of interest. Equation (7.14) shows that the value of j_o depends on the activities of the oxidised and reduced forms of the electrode system and hence in dilute solutions j_o will be low. This is one of the reasons why it is difficult to obtain accurate e.m.f. values in dilute solutions unless very rigorous precautions are taken to exclude impurities. This point has been mentioned with regard to the determination of standard electrode potentials in Chapter 6.

CONCENTRATION OVERPOTENTIAL

The origin of concentration overpotential is probably best illustrated by considering the electrolysis of a metal salt between electrodes of the same metal, for example that of silver nitrate between silver electrodes.

Before electrolysis commences, the silver electrodes will come into equilibrium with the solution and adopt the appropriate reversible potential, depending upon the activity of the silver ions in the solution. As both electrodes dip into the same solution, they will have the same reversible electrode potential and hence there will be no potential difference between them. If a small external e.m.f. is applied across the electrodes, the silver electrode connected to the negative source behaves as a cathode and that connected to the positive source as an anode. At the cathode, silver ions will be removed from solution as they are discharged to form metallic silver. At the anode, the concentration of silver ions will be increased by the dissolution of the electrode. The solutions in the immediate vicinity of the electrodes differ in concentration from the bulk of the solution: that of the anolyte is higher, of the catholyte, lower. The higher concentration of silver ions around the anode is partially reduced by migration of silver ions towards the cathode and diffusion into the bulk of the solution. Similarly, the lower concentration of silver ions around the cathode is partially increased by migration of silver ions from the anode and diffusion from the bulk solution. If, however, these transport phenomena are slow, a finite difference in concentration of silver ions around the anode and the cathode will result. This will give rise to a concentration cell with an e.m.f. which opposes the applied e.m.f. The difference between the actual operating potential of either of the electrodes and the potential which they had when in equilibrium with silver ions of the same activity as in the bulk of the solution, is the concentration overpotential of the electrode. The e.m.f. of the concentration cell produced by electrolysis is thus the sum of the anodic and cathodic overpotentials, and the applied e.m.f. must exceed the back e.m.f. of the concentration cell for electrolysis to continue.

We should now turn our attention to the situation obtaining at one of the electrodes. Before any external potential is applied to the electrodes the concentration of silver ions will be uniform throughout the solution, both electrodes being in equilibrium at a potential given by eqn. (4.5) If the uniform concentration of silver ions is c_s, and concentration is used instead of activity in eqn. (4.5), it takes the form

$$E = E^\oplus + \frac{RT}{F} \ln c_s \qquad (7.32)$$

Suppose now that we consider the electrode to which a potential E', less than E, is applied. This electrode will behave as a cathode and silver ions will be deposited from solution. For simplicity let us assume that this deposition process is so rapid that equilibrium is achieved at the electrode surface. In other words we assume that the concentration of silver ions at the electrode surface will be rapidly adjusted by the deposition reaction until it corresponds to E', this correspondence once again being given by eqn. (4.5). If the concentration of ions at the electrode surface is c_e, the appropriate relation is

$$R' = E^{\ominus} + \frac{RT}{F} \ln c_e \qquad (7.33)$$

As $E' < E$, then $c_e < c_s$ and a concentration gradient is established in the solution which will cause the diffusion of silver ions towards the electrode from the bulk of the solution. Silver ions which are migrating towards the cathode will encounter the concentration gradient and will thus be accelerated, each silver ion having a diffusion component of velocity superimposed on the migration component. The rate of arrival of silver ions at the electrode will be governed by the speed of migration and the magnitude of the concentration gradient. As soon as silver ions arrive at the electrode, they will be deposited by the fast electrode reaction so that the concentration at the electrode surface is maintained at c_e corresponding to the applied potential E'. The rate determining step in the overall process is thus the mass transport process by which ions arrive at the electrode. In an unstirred solution the concentration profile of silver ions will be a function of distance, x, from the electrode and time, t. When $t = 0$ the concentration is uniform throughout the solution at the value c_s. When the potential is applied, the concentration at the electrode surface is rapidly reduced to c_e and the solution in the neighbourhood of the electrode becomes progressively more depleted in silver ions. The situation is represented in *Figure 38* where the concentration of silver ions is shown as a function of distance from the electrode at increasing time intervals. It can be seen from *Figure 38* that in an *unstirred* solution the concentration gradient decreases with time and the diffusion process will become slower and slower.

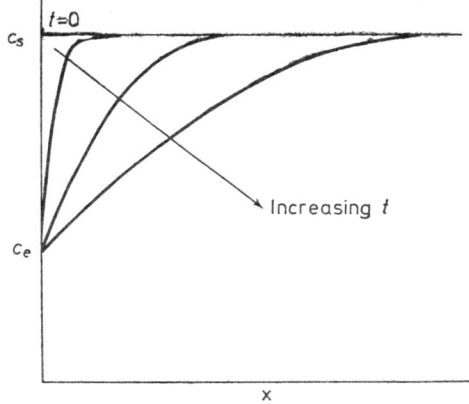

Figure 38. Variation of concentration gradient in unstirred solutions

With *stirred* solutions the situation is somewhat simpler. It may be assumed that there is a stagnant layer of solution close to the electrode in which there is no movement. In the rest of the solution the concentration of ions is maintained uniform throughout so that the concentration gradient is confined to the stagnant layer of solution which is called the *diffusion layer*. Under these conditions the concentration profile of the ions quickly reaches a steady state as depicted in *Figure 39*. A silver electrode located in the bulk of the solution would adopt a potential E, given by eqn. (7.32) and the working potential of the cathode E', is given by eqn. (7.33). The difference between these two potentials is the concentration overpotential which is given by eqn. (7.3) as

$$\eta = E' - E$$

Hence

$$\eta = \frac{RT}{F} \ln \frac{c_e}{c_s}$$

or, in general

$$\eta = \frac{RT}{zF} \ln \frac{c_e}{c_s} \tag{7.34}$$

213

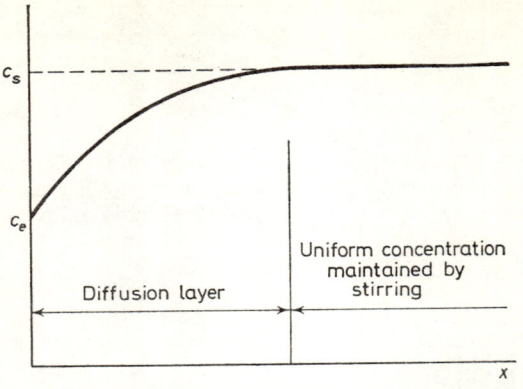

Figure 39. Steady state concentration gradient in stirred solutions

In the particular case considered above $c_e < c_s$ and η will thus be negative showing that the electrode is behaving as a cathode.

Before considering eqn. (7.34) further, we must examine the problem of mass transport in greater detail.

TRANSPORT IN ELECTROLYSIS

In electrolysis it is important to distinguish between the ions carrying the current and those which are discharged at the electrodes. In the bulk of the solution, *all* the ions present share in carrying the current according to their transport numbers. At the electrodes, the ions which are discharged are determined by the potentials of the electrodes.

Consider the electrolysis of 0.1 mol dm^{-3} $AgNO_3$ in neutral solution between silver electrodes. If a small e.m.f. is applied across the electrodes, the one connected to the negative source will behave as a cathode and the other as an anode. Silver ions will be discharged at the cathode and be deposited as metallic silver. The anode will dissolve to form silver ions in solution. Hydrogen ions and hydroxide ions from the water will also be present, in concentrations of 10^{-7} mol dm^{-3}.

214

In the bulk of the solution, the current will be carried mostly by the silver ions and nitrate ions. The contribution of the hydrogen and hydroxide ions to carrying the current will be negligible owing to their very low concentration. In the solution, the transport number of the silver ion is 0·47 and that of the nitrate ion, 0·53. Thus, for every 96 487 C of electricity passed, 0·47 mol Ag^+ migrate towards the cathode and 0·53 mol NO_3^- towards the anode. At the cathode, 1 mol Ag^+ is removed from solution as metallic silver, but only 0·47 mol Ag^+ have arrived by migration. The concentration of silver ions at the cathode surface will thus be less than in the bulk of the solution, and, in a stirred system, across a thin layer of solution adjacent to the electrode surface a concentration gradient will be set up. Eventually a stationary state will be achieved where the rate of discharge of silver ions is equal to their rate of arrival at the cathode by migration and diffusion. The situation existing during electrolysis is represented diagrammatically in *Figure 40*.

As one mol Ag^+ is discharged for every 96 487 C and only 0·47 mol arrive by migration, then 0·53 mol must arrive by diffusion. The situation at the anode is exactly the converse. One mol Ag^+ is formed by dissolution, 0·47 mol migrate away and hence 0·53 mol must diffuse away.

Consider now the nitrate ions. For 96 487 C, 0·53 mol NO_3^- migrate away from the cathode. The concentration of nitrate ions at the cathode is thus depleted, only to be made up by diffusion from the bulk of the solution of 0·53 mol. Once again, at the other electrode the converse is true: 0·53 mol NO_3^- arrive at the anode, they cannot be discharged and a concentration gradient is established causing them to diffuse into the bulk solution.

Suppose that the current passing through the solution is I. The currents carried by the migration of the silver ions and nitrate ions are 0·47 I and 0·53 I, respectively. These are the *migration currents* of the ions, and they are the only currents operative in the bulk of the solution. In the diffusion layers, however, some current is carried by the diffusion of the ions. These currents are called *diffusion currents*. Consider the cathodic diffusion layer in *Figure 40*. The nitrate migration current away from the cathode is 0·53 I. The nitrate diffusion current towards

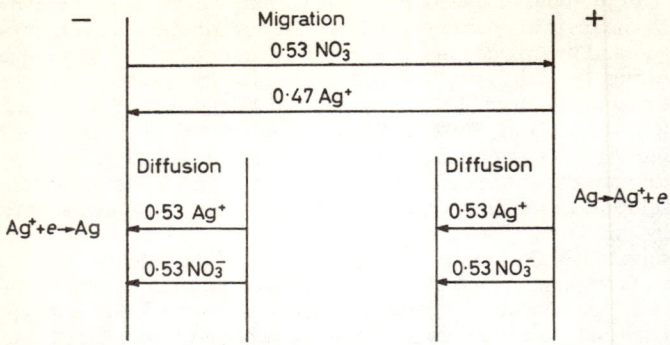

Figure 40. Electrolysis of 0.1 mol dm^{-3} AgNO$_3$

the cathode is also $0.53\ I$. The net current carried by nitrate ions in the diffusion layer must therefore be zero. The silver migration current towards the cathode is $0.47\ I$, and the silver diffusion current, also towards the cathode, is $0.53\ I$. It is thus apparent that the silver ions are responsible for carrying the whole of the current across the diffusion layer.

Similar arguments show that the silver ion carries the whole of the current across the anodic diffusion layer also. Whilst all the ions share in carrying the current in the bulk of the solution, only those which are discharged carry the current across the diffusion layers. It must be pointed out, however, that these diffusion layers are very thin, being only a fraction of a millimetre in thickness.

In the above example, the silver ion carries all the current across the diffusion layers. About half is migration current and half diffusion current. In the presence of excess of an indifferent electrolyte, however, it can be shown that most of the current carried across the diffusion layer by silver ions is diffusion current. Consider, for example, a solution which is 0.1 mol dm^{-3} with respect to AgNO$_3$ and 1 mol dm^{-3} with respect to KNO$_3$. As the silver ion is only present in low concentration compared with the other ions, its transport number will be small.

For the above solution, the transport numbers of the silver, potassium and nitrate ions are 0·04, 0·46 and 0·5 severally. The electrolysis of the solution between silver electrodes is represented in *Figure 41*. As the potassium and nitrate ions are not dis-

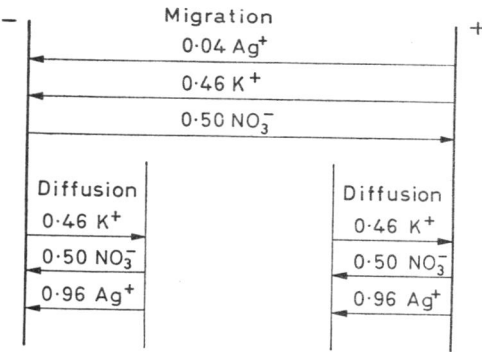

Figure 41. Electrolysis of 0·1 mol dm⁻³ $AgNO_3$ + 1·0 mol dm⁻³ KNO_3

charged, their diffusion currents must be equal and opposite to their migration currents. The silver ion must carry all the current across the diffusion layers, and as only 0·04 of the current is carried by migration, 0·96 must be carried by diffusion. This is in the presence of a tenfold excess of potassium nitrate. If the concentration of potassium nitrate were further increased or, alternatively, if that of silver nitrate were decreased, the proportion of the current carried by diffusion would be much closer to 100 per cent.

We have seen above that the only ion which takes part in the electrode reaction (the electroactive ion) carries the whole of the net current across the diffusion layers. This may be considered to consist of a diffusion current I_d and a migration current I_m. Thus, the total current I is given by

$$I = I_m + I_d \qquad (7.35)$$

217

ELEMENTARY ELECTROCHEMISTRY

The migration current is the fraction of the total current which an ion carries by migration and is thus equal to the product of the total current and the transport number of the ion in question, t_l.

$$I_m = It_l \tag{7.36}$$

The diffusion current may be deduced from Fick's first law of diffusion which states

$$\frac{dn}{dt} = -DA\frac{\partial c}{\partial x} \tag{7.37}$$

where dn/dt is the amount of substance diffusing per unit time across an area A due to a concentration gradient $\partial c/\partial x$. D is a proportionality constant called the diffusion coefficient. The minus sign in eqn. (7.37) indicates that the diffusion occurs in the opposite direction to increasing concentration gradient and partial differentials are used for the concentration gradient as strictly it is a function of time as well as distance. We shall however, only consider the concentration gradient when it has reached the steady state in a stirred solution, depicted by *Figure 39*. It may be noticed in *Figure 39* that the concentration gradient varies with distance and in order to use eqn. (7.37) to calculate the diffusion current, the approximation (originally due to Nernst) is made that the concentration gradient is linear over a distance as shown in *Figure 42*.

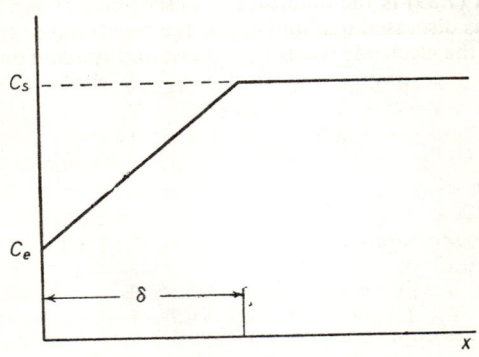

Figure 42. The Nernst approximation

Thus

$$\frac{\partial c}{\partial x} = \frac{c_s - c_e}{\delta}$$

Substituting in eqn. (7.37) but ignoring the minus sign as we are not interested in the direction of flow of the current

$$\frac{\mathrm{d}n}{\mathrm{d}t} = DA \frac{c_s - c_e}{\delta}$$

To express the rate of diffusion as a current $\mathrm{d}n/\mathrm{d}t$ must be multiplied by $z_i F$ where z_i is the charge number of the ion, thus

$$I_d = z_i FDA \frac{c_s - c_e}{\delta} \qquad (7.38)$$

Substituting from eqns. (7.36) and (7.38) into eqn. (7.35)

$$I = It_i + z_i FDA \frac{c_s - c_e}{\delta}$$

or

$$I = \frac{z_i FDA}{(1 - t_i)} \frac{c_s - c_e}{\delta} \qquad (7.39)$$

Equation (7.39) is the quantitative description of the behaviour which was discussed qualitatively at the beginning of this section in which the electrode reaction was fast and the rate determining step was the mass transport process. It shows that, as the transport number of the electroactive ion is reduced to zero by the addition of an indifferent electrolyte, the total current given by eqn. (7.39) becomes identical with the diffusion current given by eqn. (7.38).

LIMITING DIFFUSION CURRENTS

The process of diffusion in the case where an ion is being removed from solution by discharge at an electrode gives rise to an interesting phenomenon. Consider the deposition of a metal at an electrode from a stirred solution containing a large excess of an indifferent electrolyte. All the current will be carried across the diffusion layer by the electroactive ion. In the presence of the indifferent electrolyte the total current will consist entirely of diffusion current and will be given by eqn. (7.38) as

$$I = z_i FDA \frac{c_s - c_e}{\delta} \tag{7.38}$$

If the potential of the cathode is made more negative, the concentration of electroactive ions at the electrode surface will decrease. This causes the concentration gradient to become steeper so that the rate of diffusion of the electroactive ion increases and the current increases correspondingly according to eqn. (7.38). If the potential of the cathode is made progressively more negative, the current continues to increase until c_e falls to zero. Under these conditions the diffusion process has reached its maximum possible rate and the current cannot be further increased even if the cathode potential is made more negative still. This maximum value of the current is known as the *limiting diffusion current* and will be given by putting $c_e = 0$ in eqn. (7.38) so that

$$I_l = z_i FDAc_s/\delta \tag{7.40}$$

where I_l is the limiting current. It can be seen from eqn. (7.40) that the limiting diffusion current is proportional to the concentration of the electroactive ion in the bulk of the solution and this fact is used as the basis of the analytical technique of *polarography* which will be considered later.

The quantitative relation between concentration overpotential at a cathode and the current flowing in the above situation can be obtained from eqn. (7.34). A value for c_e may be obtained from eqn. (7.38) which may be rearranged to give

$$c_e = c_s - \frac{I\delta}{z_i FDA}$$

Substituting this value of c_e into eqn. (7.34)

$$\eta = \frac{RT}{zF} \ln \frac{c_s - (I\delta/z_i FDA)}{c_s}$$

or

$$\eta = \frac{RT}{zF} \ln \frac{(z_l FDAc_s/\delta) - I}{z_l FDAc_s/\delta}$$

Substituting from eqn. (7.40)

$$\eta = \frac{RT}{zF} \ln \frac{I_l - I}{I_l} \qquad (7.41)$$

where I_l is the limiting current and I is the current passing at concentration overpotential η. A plot of I against η from eqn. (7.41) is shown in *Figure 43*.

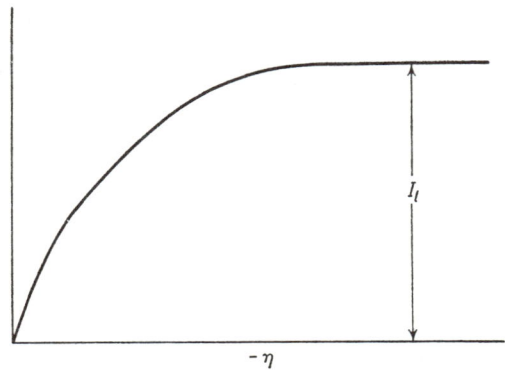

Figure 43. Variation of diffusion current with concentration overpotential

DECOMPOSITION VOLTAGES AND DISCHARGE POTENTIALS

Up to this stage we have been concerned largely with single electrodes which were in equilibrium with the solution before any polarising potential was applied to them. From a practical point of view it will be useful to consider a complete cell and we will firstly examine electrolysis in a cell where the electrodes are

221

initially in equilibrium with the solution. Consider a cell consisting of a hydrogen electrode and a chlorine electrode with hydrochloric acid as the electrolyte.

$$Pt;H_2/HCl/Cl_2;Pt$$

In this cell the chlorine electrode is found to be the positive pole and the hydrogen electrode the negative pole of the cell. If an external e.m.f. is applied to the cell such that the negative of the applied e.m.f. is connected to the hydrogen electrode and the positive to the chlorine electrode, the applied e.m.f. will oppose the cell e.m.f. When the applied e.m.f. is exactly equal to the cell e.m.f., no current will flow and the hydrogen and chlorine electrodes will be at their reversible potentials. This is the situation which obtains when the cell e.m.f. is measured with a potentiometer. If the applied e.m.f. is greater than the e.m.f. of the cell, a current will flow round the circuit, the energy being derived from the external source, the solution in the cell being electrolysed. As the applied e.m.f. is greater than the cell e.m.f., the chlorine electrode must have applied to it a potential more positive than its reversible potential. The chlorine electrode thus

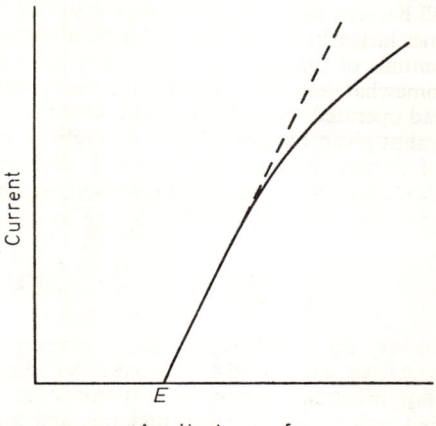

Figure 44. Variation of electrolysis current with applied e.m.f.

222

behaves as an anode and the hydrogen electrode as a cathode. The relationship between the current flowing and the applied e.m.f. is shown in *Figure 44*. Suppose that the reversible e.m.f. of the cell is E, the applied e.m.f. is E_{app} and the resistance of the circuit is R. As mentioned above, when $E_{app} = E$, the current will be zero. When $E_{app} > E$, electrolysis will occur and if the electrodes operated in a perfectly reversible manner the cell e.m.f. would remain at the value E. Under these circumstances a net e.m.f. of $(E_{app} - E)$ would drive current round a circuit of resistance R and the current I would be given by Ohm's law as

$$I = \frac{E_{app} - E}{R} \qquad (7.42)$$

Equation (7.42) shows that a plot of I against E_{app} is a straight line of slope $1/R$ and that $I = 0$ when $E_{app} = E$. This will be the case for low currents when the electrodes will operate more or less reversibly. At higher values of E_{app}, larger currents will flow and the electrodes will be subject to polarisation. As the chlorine electrode is the anode of the system, its potential will become more positive and conversely the potential of the hydrogen electrode will become more negative. As a result of this, the cell e.m.f. becomes larger by an amount corresponding to the sum of the overpotentials of the two electrodes. The net e.m.f. driving current is somewhat less than would have been the case if the electrodes had operated reversibly and as a result the current will be less than that given by eqn. (7.42). At higher currents then, the graph of current against applied e.m.f. deviates from the straight line of slope $1/R$. As the overpotentials increase with increasing currents, the deviation also increases with increasing current.

The current–voltage curve in *Figure 44* refers to a system where the electrodes are in equilibrium with the solution before any external e.m.f. is applied. In most cases of practical electrolysis however, the electrodes are not initially in equilibrium with the solution. Consider, for example, the electrolysis of hydrochloric acid between platinum electrodes. No hydrogen or chlorine gas will be present before electrolysis commences and the electrodes cannot be in equilibrium with the solution. Under these circumstances the current–voltage curve for the system takes the form

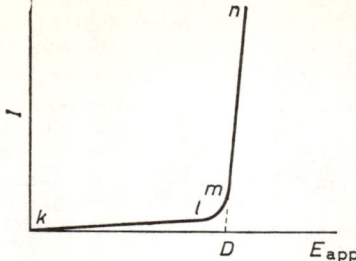

Figure 45. Decomposition curve

illustrated in *Figure 45*. When a very small e.m.f. is applied across the electrodes, hydrogen ions move towards the negative electrode (cathode) and are discharged to form hydrogen gas.

$$H^+ + e \rightarrow \tfrac{1}{2}H_2$$

Similarly, chloride ions move towards the positive electrode (anode) and are discharged to form chlorine gas.

$$Cl^- \rightarrow \tfrac{1}{2}Cl_2 + e$$

As soon as traces of these products appear at the electrodes, the system constitutes a galvanic cell with an e.m.f. which is in opposition to the applied e.m.f. The back e.m.f., E, of the galvanic cell is given by

$$E = E_{Cl_2} - E_{H_2}$$

where E_{Cl_2} and E_{H_2} are the potentials of the chlorine and hydrogen electrodes respectively. These are given by eqn. (4.1)

$$E_{Cl_2} = E_{Cl_2}{}^\circ + \frac{RT}{F} \ln \frac{p_{Cl_2}{}^{\frac{1}{2}}}{a_-}$$

$$= E_{Cl_2}{}^\circ + \frac{RT}{2F} \ln p_{Cl_2} - \frac{RT}{F} \ln a_-$$

224

where a_- is the activity of the chloride ions and p_{Cl_2} is the partial pressure of chlorine at the electrode.

$$E_{H_2} = \frac{RT}{F} \ln \frac{a_+}{p_{H_2}^{\frac{1}{2}}}$$

$$= \frac{RT}{F} \ln a_+ - \frac{RT}{2F} \ln p_{H_2}$$

where a_+ is the activity of the hydrogen ions and p_{H_2} the partial pressure of the hydrogen gas at the electrode.

The e.m.f. of the galvanic cell is thus

$$E = E_{Cl_2}^{\circ} + \frac{RT}{2F} \ln p_{H_2} p_{Cl_2} - \frac{RT}{F} \ln a_+ a_- \qquad (7.43)$$

Assuming that the activities of the ions remain constant, eqn. (7.43) shows that the back e.m.f. increases as the partial pressures of the gases produced increase. When enough products have accumulated at the electrodes to produce a back e.m.f. equal to the applied e.m.f., electrolysis should cease. Ideally there should only be a transient current when a small e.m.f. is applied to the system. This transient current should consist partly of the current required to discharge the necessary amounts of products at the electrodes and partly of the current required to form the electrical double layer. The former is called *faradaic current* as it results in electrochemical action and the latter is known as *non-faradaic* as no electrochemical change results, the current being analogous to the current required to charge a condenser.

In spite of the fact that both the faradaic and non-faradaic currents should only be transient it may be seen from *Figure 45* that there is a very small but continuous current at values of applied e.m.f. between k and l. This is called the *residual current*; it arises from the fact that the products of electrolysis diffuse away from the electrodes. The residual current is thus required to maintain the appropriate concentrations of hydrogen and chlorine at the electrodes.

As the applied e.m.f. is increased, the concentrations of hydrogen and chlorine required to oppose the applied e.m.f.

become greater, diffusion is thus more rapid and the residual current increases. There comes a stage between l and m in *Figure 45* where the partial pressures of hydrogen and chlorine at the electrodes exceed 1 atm. At this stage, the gases are evolved from the electrodes, the back e.m.f. can increase no more, and if the applied e.m.f. is further increased there will be a net e.m.f. causing free electrolysis to occur. This state of affairs corresponds to the section mn in *Figure 45* and the graph now takes the form of *Figure 44*. The section mn is usually extrapolated back to zero current to give the value of applied e.m.f., D, at which appreciable electrolysis currents can flow. The applied e.m.f. at D is known as the *decomposition voltage* but it has no theoretical significance. The concept has some use in practical electrolysis and the value D may be called the *practical decomposition voltage*.

It may be seen from *Figure 44* that it is only necessary to infinitesimally exceed the reversible e.m.f. of a cell for electrolysis to occur. The minimum value of applied e.m.f. required for electrolysis is thus equal to the reversible e.m.f. of the cell and may be called the *reversible decomposition voltage*. Under these conditions any electrolysis current flowing will be insignificant. If an appreciable electrolysis current is required, the applied e.m.f. must overcome any overpotentials associated with the electrode process and must also include an amount equal to the IR drop through the electrolyte. This last term will of course depend on the geometrical arrangement of the electrodes within the cell which illustrates the limitation of the concept of decomposition voltage.

Although the liberation of hydrogen at a cathode has been considered in the above example, a similar situation exists when a metal is being deposited. Whilst the deposited metal does not completely cover the electrode, its activity is variable and the applied e.m.f. may be successfully opposed. As soon as the electrode is completely covered, it behaves as the pure metal with an activity of unity. The back e.m.f. cannot increase beyond this stage and a further increase in the applied e.m.f. causes electrolysis to occur.

As the total e.m.f. applied to an electrolysis cell includes the IR drop through the electrolyte, which depends on the geometry of the cell, it is more significant to consider the potentials of the individual electrodes. The operating potentials of the electrodes will be equal to their reversible potentials plus any overpotential and these potentials may be called *discharge potentials*. When the processes are the deposition or dissolution of a metal, the potentials are sometimes called deposition or dissolution potentials. It is apparent that the minimum discharge potentials are the reversible potentials, but for the passage of significant currents the necessary overpotential terms must be taken into account.

In the study of electrode processes then, it is more usual to plot current against the potential of the electrode of interest rather than a graph of current against total applied voltage as in *Figure 45*. This practice is followed in polarography which is the analytical technique mentioned earlier in connection with concentration overpotential.

POLAROGRAPHY

In polarography, the electrolysis is usually carried out between mercury electrodes when the interest lies in cathodic processes. The anode is a pool of mercury of large surface area, so that the anodic current density will be low and the anode kept free from polarisation. The cathode is a small drop of mercury which hangs from the tip of a capillary tube connected to a reservoir of mercury. By adjusting the feed from the reservoir, the drop of mercury can be made to slowly increase in size until it finally drops away and a new drop then starts to form at the end of the capillary. In this way, the cathode surface is continually renewed, with a new drop being formed about every three to five seconds. The cathode is thus kept clean and free from contamination by any products of electrolysis. The apparatus is usually arranged so that the mercury drops which have previously constituted the cathode, fall into the mercury pool anode which is located in the bottom of the electrolysis cell. A reference electrode may be placed in the solution so that the operating potential of the cathode can be determined.

The polarographic technique is based on concentration polarisation, and the associated fact, that limiting diffusion current is proportional to the concentration of the electroactive species in the bulk of the solution. Suppose a solution of metal ions were to be examined polarographically. The ions will discharge at the mercury cathode to form amalgam at the appropriate discharge potential. In order that the current carried across the diffusion layer by the discharging ions shall be mostly diffusion current, excess of an indifferent electrolyte (frequently potassium chloride) is added to the solution and an increasingly negative potential is applied to the cathode.

At low applied potential, a residual current is observed but when the cathode potential becomes sufficiently negative the metal ions will be discharged to form metal amalgam in the mercury cathode. The drops of amalgam from the cathode fall into the anode pool and the anodic process may become the dissolution of the metal from the amalgam to form metal ions in solution again. As the cathode potential is made progressively more negative, the current increases until the diffusion current of the metal ions reaches the limiting value appropriate to the particular concentration of metal ions in the solution. If the total current through the cell were entirely due to the metal ion diffusion current, the cell current would remain constant at this stage despite increasingly negative applied potential. In practice the cell current continues to vary with applied potential owing to the presence of residual and other spurious currents. This variation is only slight and is illustrated in *Figure 46* where the variation of current through the cell with increasingly negative cathode potential is plotted. Such a diagram is called a *polarogram* which is, of course, similar in form to *Figure 43* which shows variation of diffusion current with cathodic overpotential.

As the limiting diffusion current of an ion is proportional to its concentration in solution, the interval I_l on the polarogram is a measure of the concentration of metal ions in solution. In order to determine the actual concentration of the ions, the apparatus must first be calibrated with solutions of known concentration. Polarography thus provides a method of quantitative analysis.

To arrive at the equation for the *polarographic wave* shown in *Figure 46* it will be useful to consider the general case of reduction at a cathode where the oxidised form of the system diffuses

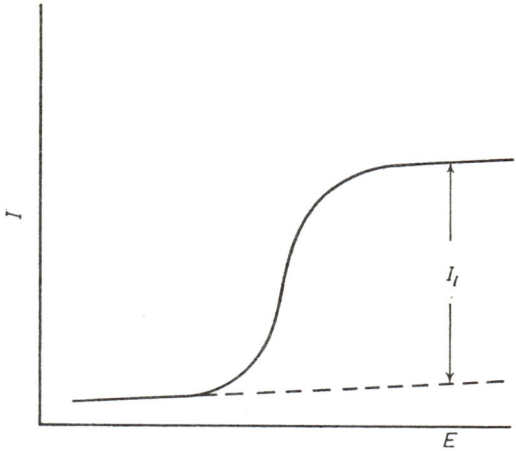

Figure 46. A polarogram

towards the cathode to become the reduced form as a result of the electrode reaction.

$$Ox + ze \rightarrow Red$$

It has been shown earlier that for reduction at a cathode in which the rate determining step is the mass transport process the current is proportional to the rate of diffusion of the oxidised form to the electrode. This in turn depends upon the difference in the concentrations of the oxidised form at the electrode surface and in the bulk of the solution. Thus

$$I = k_O([Ox]_s - [Ox]_e) \qquad (7.44)$$

where $[Ox]_s$ and $[Ox]_e$ are the concentrations of the oxidised form in the bulk of the solution and at the electrode surface respectively. k_O is a proportionality constant which will involve the diffusion coefficient of the oxidised form. The limiting diffusion current, I_l, is achieved when $[Ox]_e = 0$, hence

$$I_l = k_O[\text{Ox}]_s \qquad (7.45)$$

Substituting from eqn. (7.45) into eqn. (7.44)

$$I = I_l - k_O[\text{Ox}]_e$$

or

$$[\text{Ox}]_e = \frac{I_l - I}{k_O} \qquad (7.46)$$

If the reduced form of the system is soluble in mercury or water, it will diffuse away from the electrode surface once it has been formed. There will be a stationary concentration of the reduced form at the electrode surface which will be governed by its rate of production and its rate of disappearance by diffusion. If the electrode reaction itself is fast, the rate of production of the reduced form is determined by the rate of arrival of the oxidised form which determines the current. The rate of diffusion away from the electrode surface of the reduced form will depend only on its concentration at the electrode surface as its concentration remote from this point will be virtually zero. Equating the rate of production and the rate of disappearance at the electrode surface of the reduced form we have

$$I = k_R[\text{Red}]_e \qquad (7.47)$$

where $[\text{Red}]_e$ is the concentration of the reduced form at the electrode surface and k_R is a constant which will include the diffusion coefficient of the reduced form. Rearranging eqn. (7.47),

$$[\text{Red}]_e = \frac{I}{k_R} \qquad (7.48)$$

If the electrode reaction is fast, as has been assumed above, there will be no activation polarisation and the potential of the electrode will be related to the activities of the oxidised and reduced forms at the electrode surface by eqn. (4.1)

$$E = E^\circ + \frac{RT}{zF} \ln \frac{a_{\text{Ox}}}{a_{\text{Red}}} \qquad (4.1)$$

which may be re-written as

$$E = E^{\circ} + \frac{RT}{zF} \ln \frac{y_{Ox}[Ox]_e}{y_{Red}[Red]_e} \qquad (7.49)$$

where y_{Ox} and y_{Red} are the activity coefficients of the oxidised and reduced forms respectively.

Substituting from eqns. (7.46) and (7.48) into eqn. (7.49)

$$E = E^{\circ} + \frac{RT}{zF} \ln \frac{y_{Ox}}{y_{Red}} \cdot \frac{k_R}{k_O} \frac{I_l - I}{I}$$

or

$$E = E^{\circ} + \frac{RT}{zF} \ln \frac{y_{Ox}}{y_{Red}} \frac{k_R}{k_O} + \frac{RT}{zF} \ln \frac{I_l - I}{I} \qquad (7.50)$$

The values of the constants k_O and k_R in eqn. (7.50) depend upon the number of electrons involved in the electrode reaction, the mass of the mercury drops, the lifetime of the mercury drops and the diffusion coefficient of the species concerned. All these factors except the last are common for both the oxidised and reduced forms and the equation which relates current to concentration for a dropping mercury electrode was originally deduced by Ilkovic who showed that the constants k were dependent upon the square root of the diffusion coefficients. Thus

$$k_R/k_O = D_R^{\frac{1}{2}}/D_O^{\frac{1}{2}}$$

where D_R and D_O are the diffusion coefficients of the reduced and oxidised forms. Equation (7.50) may thus be written in the form

$$E = E^{\circ} + \frac{RT}{zF} \ln \frac{y_{Ox}}{y_{Red}} \frac{D_R^{\frac{1}{2}}}{D_O^{\frac{1}{2}}} + \frac{RT}{zF} \ln \frac{I_l - I}{I} \qquad (7.51)$$

The activity coefficients y_{Ox} and y_{Red} can be regarded as independent of the concentrations of the oxidised and reduced forms since there will be a large excess of an indifferent electrolyte present in the solution. The second term on the right-hand side of eqn. (7.51) may thus be regarded as a constant. When $I = \frac{1}{2}I_l$

the third term becomes zero and writing the electrode potential at this point as $E_{\frac{1}{2}}$,

$$E_{\frac{1}{2}} = E^{\circ} + \frac{RT}{zF} \ln \frac{y_{Ox}}{y_{Red}} \frac{D_R^{\frac{1}{2}}}{D_O^{\frac{1}{2}}} \qquad (7.52)$$

As the right-hand side of eqn. (7.52) is composed entirely of constants at constant temperature for a particular electrode process, $E_{\frac{1}{2}}$ must also be a constant under these conditions. As $E_{\frac{1}{2}}$ is the electrode potential when $I = \frac{1}{2}I_l$ it is called the *half-wave potential* and eqn. (7.51) may be written

$$E = E_{\frac{1}{2}} + \frac{RT}{zF} \ln \frac{I_l - I}{I} \qquad (7.53)$$

This is the equation of the polarographic wave shown in *Figure 46* and $E_{\frac{1}{2}}$ is the potential of the electrode at the point of inflexion of the curve.

The potential at which the rate of the cathodic process becomes appreciable (the discharge potential) varies with the concentration of the electroactive species in solution and cannot be regarded as characteristic of a particular species. It will be appreciated from eqn. (7.52) however, that half-wave potentials are independent of concentration and are, in fact, characteristic of the particular species involved in the electrode reaction.

Polarography is thus a method of both quantitative and qualitative analysis. It must be pointed out, however, that although half-wave potentials are independent of the concentration of the ion to which they refer, they do depend upon the concentration and nature of the supporting electrolyte. When reference is made to tables of half-wave potentials in order to identify a species being analysed, the potentials appropriate to the supporting electrolyte in the experiment must be used.

The above treatment has been restricted to the case where mass transport is the rate determining step, the actual electrode process being much faster so that the electrode may be considered to be in equilibrium with the oxidised and reduced forms of the system at the electrode surface. Polarographic waves obtained under these conditions are known as *reversible waves*. If there is no great difference in the speeds of the mass transport process and the electrode process so that neither is unequivocally the rate determining step, then some activation polarisation may occur at the electrode and the treatment becomes rather more complex. The polarographic waves for these conditions are known as *irreversible waves*. There is still a characteristic half-wave potential for an irreversible system but in addition to being dependent on the diffusion coefficients it depends also on the lifetime of the mercury drop, the transfer coefficient, α, for the electrode process and the rate constant k_1^0, for the cathodic process in the absence of any potential. It will be recalled that these latter two factors occurred in the treatment of activation overpotential which was discussed earlier in this Chapter.

For a reversible polarographic wave, eqn. (7.53) shows that a plot of E against $\ln[(I_l - I)/I]$ should give a straight line of intercept $E_{\frac{1}{2}}$ and slope RT/zF. In this way the half-wave potential and the number of electrons involved in the electrode reaction may be determined.

Polarographic examination may be applied to solutions containing more than one component. Suppose a solution containing zinc and cadmium ions were to be examined polarographically. These ions will discharge at a mercury cathode to form amalgams at the appropriate discharge potentials. So that these electro-active ions shall arrive at the electrode by diffusion only, an excess of an indifferent electrolyte is added to the solution and an increasingly negative potential is applied to the cathode.

At low applied potential a residual current is observed, but when the cathode potential becomes sufficiently negative, one of the cations will be discharged. In the present example, cadmium ions have a less negative discharge potential than zinc ions and the first cathodic process to occur will be the discharge of cadmium ions to form cadmium amalgam in the mercury electrode. As the cathode potential becomes more negative than the discharge potential of the cadmium ions, the current increases rapidly with increasing applied e.m.f. until the diffusion current of the cadmium ions reaches the limiting value appropriate to the particular concentration of cadmium ions in the solution. After this point the current increases only slowly, this increase being entirely due to the increase in the residual current.

The current will continue to vary very slowly until the potential of the cathode becomes more negative than the discharge potential of the zinc ions, when the current will again increase sharply due to the discharge of zinc ions. The rapid rise in current continues until the diffusion current of the zinc ions reaches its limiting value when the rate of increase of current with potential falls off. At this point, the total current is the sum of the diffusion currents of both the zinc and cadmium ions together with the residual and other spurious currents. The resultant polarogram is shown in *Figure 47* which shows the half-wave potentials and

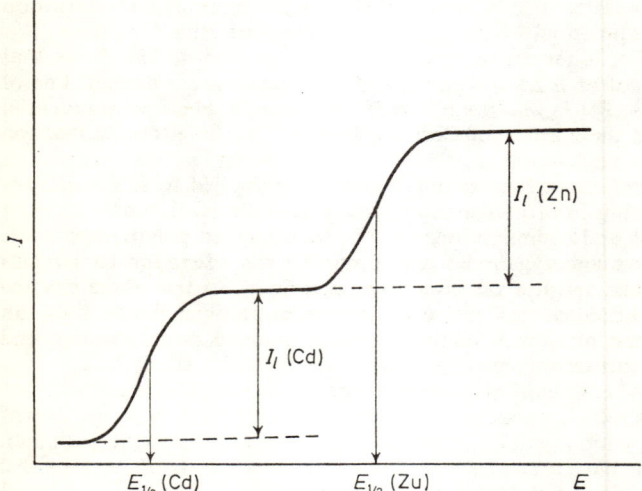

Figure 47. Polarogram of two components

limiting diffusion currents of the two components. The latter are usually measured as the perpendicular distance to the graph through the points of inflexion but have been drawn in a different position in *Figure 47* for the sake of clarity.

The range of potentials over which a dropping mercury electrode may be used is limited at the negative end of the range by the cathodic discharge of hydrogen from the water in which the solution is made up and at the positive end of the range by the anodic dissolution of the mercury electrode itself. The range of potential thus lies between about $+0.25$ V (dissolution of mercury) and -1.6 V (hydrogen discharge in acid solutions) to -2.0 V (hydrogen discharge in alkaline solutions). If polarograms are extended to these negative values of cathode potential, a hydrogen discharge wave will be observed.

Analyses may be carried out on very small samples (e.g. 2 cm^3) and down to concentrations of 10^{-5} mol dm^{-3} or 10^{-6} mol dm^{-3}. Any ionic or molecular species which is capable of reduction at a cathode can be analysed polarographically with a dropping mercury cathode. It is thus possible to analyse many organic systems with this technique.

In practice, it is usually advisable to remove dissolved oxygen from the solutions by bubbling nitrogen through them both prior to and during electrolysis. The presence of oxygen can generate large polarographic waves, corresponding to its reduction to hydrogen peroxide or water, which often mask the waves of other substances being analysed.

It is also possible to carry out anodic polarography, but in this case mercury anodes are unsuitable, as they tend to go into solution at low positive potentials. In anodic polarography the polarised electrode is usually a rotating platinum anode. Systems which are capable of being oxidised at an anode may be analysed by this technique.

AMPEROMETRIC TITRATIONS

(a) One polarised electrode

In polarography the measurement of a limiting diffusion current may be used to determine the concentration of an electroactive substance. The change in the concentration of such a substance during a titration may thus be followed by measuring the limiting diffusion current of the substance in a polarographic cell. The end-point of the titration may be obtained from a plot of current against the volume of titrant added. As the technique is one of current measurement, such titrations are known as *amperometric titrations with one polarised electrode* since, in polarography, one electrode (the mercury drop) is polarised and the other electrode is unpolarised. The form of the curve obtained by plotting current against volume of titrant depends upon the potential applied to the mercury drop and also upon the form of the polarograms of the two systems involved in the titration.

If two substances A and B can be polarographically reduced and they also form a precipitate, then they can be titrated against one another amperometrically. Suppose that the half-wave potential of B is more negative than that of A so that the

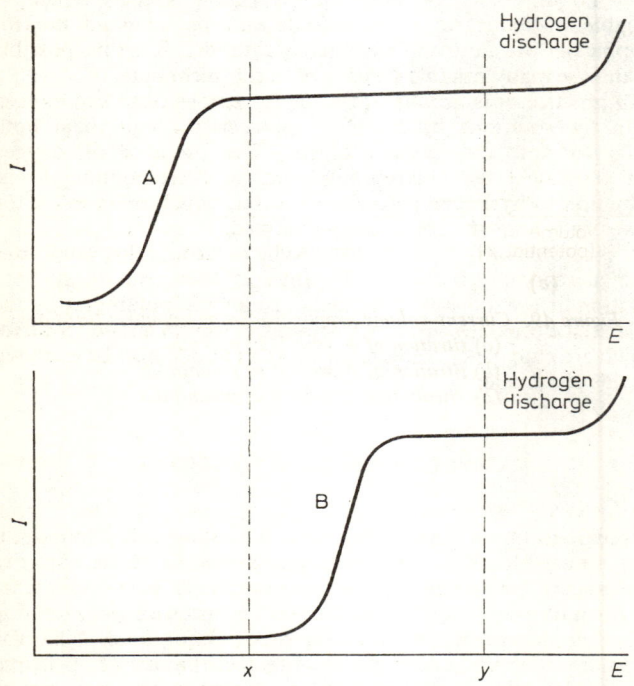

Figure 48. Polarograms of substances A and B

individual polarograms are of the form represented in *Figure 48*. Suppose that A is placed in the polarographic cell and that B is to be added from a burette. If the cathode potential is fixed at *x*, a current will be observed initially due to the reduction of A at the electrode. As B is added, the current will decrease as the concentration of A decreases until all the A has been precipitated when the only current will be the residual current. Addition of excess of B beyond the end-point will have no effect on the current as B is not reduced at the potential *x*. The form of the current–volume graph will thus be as illustrated in *Figure 49(a)*. If the

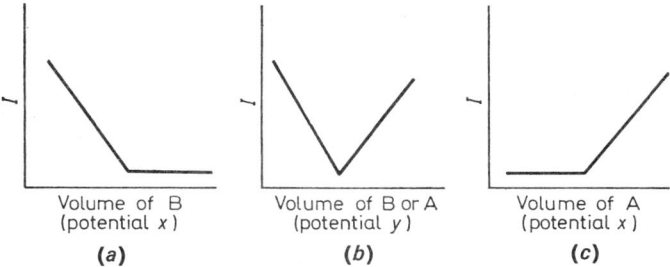

Volume of B	Volume of B or A	Volume of A
(potential *x*)	(potential *y*)	(potential *x*)
(*a*)	(*b*)	(*c*)

Figure 49. Current–volume graphs from amperometric titrations
(a) titration of A with B at potential x
(b) titration of A with B at potential y
(c) titration of B with A at potential x

cathode potential is fixed at *y*, the current would decrease initially in the same way as B was added but after the end-point the excess B will be reduced at the potential *y* and the current will increase after the end-point due to the increasing amount of B in solution. The current–volume graph thus takes the form shown in *Figure 49(b)*. This would also be the form of the graph if B were originally placed in the polarographic cell with A in the burette, provided that the cathode potential was fixed at *y*. With B in the cell, A in the burette and the cathode potential at *x* it will be appreciated that the current–volume graph takes the form of *Figure 49(c)*).

(b) Two polarised electrodes

Amperometric titrations with one polarised electrode are related to polarography which is dependent upon concentration polarisation. There is another type of amperometric titration which is dependent upon activation polarisation and which is known as an *amperometric titration with two polarised electrodes*. The classical example of such titrations is the titration of iodine with thiosulphate.

$$I_2 + 2S_2O_3^{2-} \rightarrow 2I^- + S_4O_6^{2-}$$

The iodine, dissolved in potassium iodide solution, is placed in a cell fitted with two identical platinum electrodes. A small e.m.f. is applied across the electrodes so that the cell passes a small current. The magnitude of this current will depend on the e.m.f. applied to the cell and the form of the current–overpotential graph for the I_2/I^- system. This system has a high exchange current density and thus the form of the appropriate graph will be as illustrated in *Figure 50(a)* [cf. *Fig. 37*]. As both electrodes are the same size the current density at each will be the same. If the applied e.m.f. is E the situation may be represented on

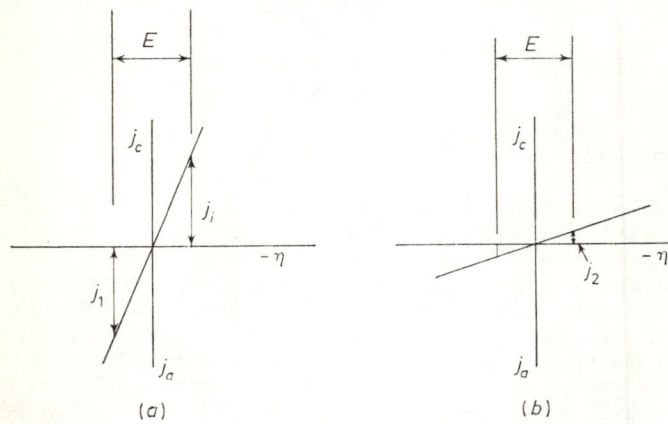

Figure 50. Current density–overpotential graphs
        (a) I^-/I_2 (b) $S_2O_3^{2-}/S_4O_6^{2-}$

238

Figure 50(a) by marking off an interval E on the overpotential axis and adjusting the position of this interval to give equal anodic and cathodic current densities. It will be seen from the diagram that this corresponds to a current density j_1 at each electrode.

As thiosulphate is added from a burette, the concentration of iodine is reduced but as long as there is some iodine and iodide present the anodic and cathodic reactions can be maintained. When the concentration of iodine becomes very low, there is some reduction of the current due to concentration polarisation. After the end-point there will be no iodine present and the only redox couple which can take part in the electrode reactions is the thiosulphate/tetrathionate couple which is highly irreversible and thus has a very low exchange current density. The current–overpotential graph for this system will thus be as shown in *Figure 50(b)* [once again cf. *Fig. 37*]. It will be appreciated that the application of the e.m.f. E to the $S_2O_3^{2-}/S_4O_6^{2-}$ system will result in a current density j_2 which is negligible compared with j_1. If current is plotted against the volume of thiosulphate added in this titration the graph takes the form of *Figure 51* and the end-point can be readily determined. In this application the

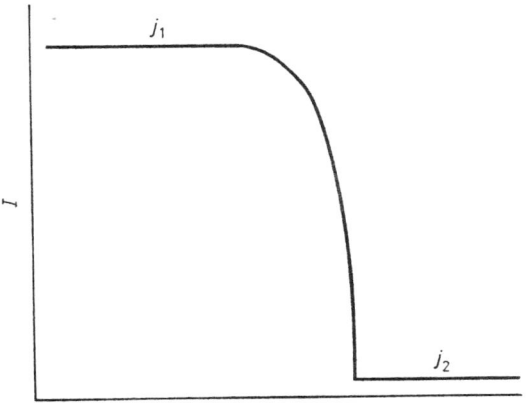

Volume of thiosulphate

Figure 51. Amperometric titration of iodine with thiosulphate

239

technique is known as the dead stop end-point technique as the current drops fairly rapidly, virtually to zero at the end-point. Amperometric titrations with two polarised electrodes thus depend on the relative reversibilities of the two systems involved.

FURTHER READING

The following list contains some suggestions for further reading which will provide an expansion and extension of the material covered in this book. The list is not comprehensive but the books quoted will give further references to more specialised topics.

Basic texts

C. W. Davies, *Electrochemistry*, Newnes, London, 1967.

S. Glasstone, *Introduction to Electrochemistry*, van Nostrand, New York, 1942.

V. Gold, *pH Measurements*, Methuen, London, 1956.

L. L. Leveson, *Introduction to Electroanalysis*, Butterworths, London, 1964.

J. J. Lingane, *Electroanalytical Chemistry*, 2nd edition, Interscience, New York, 1958.

G. R. Palin, *Electrochemistry for Technologists*, Pergamon Press, 1969.

E. C. Potter, *Electrochemistry—Principles and Applications*, Cleaver–Hume Press, London, 1956.

Advanced Texts

A. J. Bard, ed., *Electroanalytical Chemistry*, Vols 1 and 2, Edward Arnold, London, 1966.

B. E. Conway, *Theory and Principles of Electrode Processes*, Ronald Press, New York, 1965.

C. W. Davies, *Ion Association*, Butterworths, London, 1962.

P. Delahay, *New Instrumental Methods in Electrochemistry*, Interscience, New York, 1954.

P. Delahay, *Double Layer and Electrode Processes*, Interscience, New York, 1965.

H. S. Harned and B. B. Owen, *The Physical Chemistry of Electrolyte Solutions*, 3rd edition, Reinhold, New York, 1958.

G. Kortüm, *Treatise on Electrochemistry*, 2nd edition, Elsevier, London, 1965.

G. Millazo, *Electrochemistry*, Elsevier, London, 1963.

C. B. Monk, *Electrolytic Dissociation*, Academic Press, London, 1961.

R. A. Robinson and R. H. Stokes, *Electrolyte Solutions*, 2nd edition, Butterworths, London, 1959.

Books containing electrochemical problems in terms of SI units

H. E. Avery and D. J. Shaw, *Basic Physical Chemistry Calculations*, Butterworths, London, 1971.

H. E. Avery and D. J. Shaw, *Advanced Physical Chemistry Calculations*, Butterworths, London, 1971.

INDEX

241

INDEX

243

INDEX